A PENGUIN SPECIAL

Gluttons for Punishment

James Erlichman was born in New York City in 1949. In 1966 he spent a year as an exchange student at Tonbridge School in Kent but returned to America to take an honours degree in American history and politics from Brown University. In 1971 he returned to Britain where he took a Graduate Diploma in Historical Studies at Cambridge University before deciding to become a journalist. Prior to joining the *Guardian* in 1979 he worked in the provincial press and then on Fleet Street as a reporter for the London *Evening News*. In 1983 he was voted Industrial Feature Writer of the year in the Blue Circle journalism awards. Originally on the financial staff of the *Guardian*, he is now Chemicals Correspondent.

James Erlichman lives in London with his wife and two children. He enjoys cooking in his spare time.

James Erlichman

Gluttons for Punishment

Penguin Books

Penguin Books Ltd, Harmondsworth, Middlesex, England
Viking Penguin Inc., 40 West 23rd Street, New York, New York 10010, U.S.A.
Penguin Books Australia Ltd, Ringwood, Victoria, Australia
Penguin Books Canada Limited, 2801 John Street, Markham, Ontario, Canada L3R 1B4
Penguin Books (N.Z.) Ltd, 182–190 Wairau Road, Auckland 10, New Zealand

Published in Penguin Books 1986

Made and printed in Great Britain by
Richard Clay (The Chaucer Press) Ltd, Bungay, Suffolk
Filmset in Monophoto Plantin

To Susie

Contents

Acknowledgements **9**

Prologue **11**

1/ For What We Are About to Receive **13**

2/ The Guinea-Pig Generation **28**

3/ The Human Tragedies **50**

4/ Down on the Farm: How Chemicals and Drugs Fuel the Food Machine **67**

5/ In the Boardroom: The Abuse of Power by the Chemical and Drug Companies **81**

6/ The Government Watchdog: Ferocious or Toothless? **95**

7/ The Right to Know and the Right to Choose: Critics and Alternatives **122**

Conclusion **144**

Index **149**

Acknowledgements

Few people have the courage and vision to examine a vested interest with an open mind. To a rare handful of such brave souls I owe a considerable debt. They include Ian Dalzell of the National Farmers' Union and Chris Major of ICI. The others know who they are, but have asked that I do not say. The candour of all of them was invaluable.

I am also grateful to Alastair Hay, Professor Alan Linton and Dr Bernard Rowe for their endless patience in explaining to me the rudiments of their science. Chris Rose from Friends of the Earth, Dr Alan Long of the Vegetarian Society and Peter Snell of the London Food Commission were all admirable advocates of their points of view. Diane Montague and John Malcolm both helped me to try to unlock the complexities of contract farming in Britain. Gordon Applebe and Joy Wingfield of the Pharmaceutical Society gave both their time and insight and Tony Venables of BEUC in Brussels gave me essential access to EEC documents.

Both Maurice Frankel and my colleague at the *Guardian*, Richard Norton-Taylor were kind enough to read and comment upon the manuscript. But I shoulder unreservedly the errors of fact and judgement which may remain.

Finally, I need to thank my wife Susie who was both an uncompromising critic and a tolerant guide.

Prologue

This is a book about greed and addiction. A lot of people make money when our crops are dosed with excessive pesticides and our animals are given too many antibiotics and hormones. Many farmers are locked into a treadmill of chemical agriculture and cannot escape. Too many veterinarians, faced with the need to survive, sign prescriptions when they should not. The chemical industry, which chafes against restrictions on its human drug sales, is having a comparative field-day selling veterinary medicines and pesticides. The big food companies and supermarkets deserve a share of the blame. The demands they make for cosmetically perfect produce and lean meat at the lowest price force farmers to resort to the medicine kit and the pesticide spray.

But before we start pointing fingers at too many people, we ought to look at ourselves. As consumers we too have become unwitting addicts to chemical agriculture. We have come to expect cheaper, more attractive-looking food. We grumble, but on the whole, we get what we pay for, thanks to the growth-boosting properties of pesticides, antibiotics and hormones. In real terms fresh foods have never been cheaper or apparently more appealing. We now choose from rows of unblemished, polythene-wrapped chickens at reasonable prices, stacks of lean minced beef, and lettuces so pristine in appearance that their acquaintance with insects, or even soil, seems remote. The high speed agriculture machine which churns out these products simply could not function without drugs and pesticides.

The government and its eminent scientists insist that we are safe in their hands because all the drugs and pesticides approved for sale have been thoroughly tested.

Alarming incidents have already occurred which warn that all may not be well. European outrage erupted when it was discovered in Italy in 1980 that veal baby food had been heavily contaminated with high concentrations of hormones which had been injected into calves. Some Italian infants who ate food from the contaminated jars developed breasts and enlarged genitals. Intense consumer protest on the continent finally forced the European Commission to support a total ban on all hormone implants in October 1985. Despite implacable opposition from the British government which defends the safety of hormone implants, the Council of Ministers decreed a ban on 21 December 1985. But it will not become law until January 1988 at the earliest and Britain has been given special exemption, and will be allowed to use hormone implants for a further year, until January 1989. So for the time being implants are legal and their residues remain in the beef and veal we eat.

Although the hormone controversy has dominated the headlines, other drugs constantly fed to meat animals are likely to prove a bigger hazard. In Britain and in the United States there is growing evidence that drug-resistant strains of salmonella and other bacteria, which have been spawned by excessive feeding of antibiotics to livestock, have spread to the human population.

Vegetarians should not feel complacent. A recent British survey revealed that a third of all fresh fruits and vegetables are 'contaminated' with detectable levels of pesticides – including DDT and other officially banned and highly toxic substances.

Has greed got the better of all of us, then? And have the authorities in whom we trust failed to monitor abuses and prosecute the culprits? Some medical experts believe that the chemical plague we have created is already causing unexplained illness, chronic disease and even death. Consumer and environmental groups also suspect that a hidden alliance exists between the powerful farming lobby, the giant chemical and drug companies and the governments, which has led us down the garden path to food which is increasingly dangerous to eat.

This book tries to walk the tightrope between fanatical opinion and smug complacency. It seeks to assess the chemical dangers we face, to explain why the use of pesticides, hormones and antibiotics is burgeoning and to ask whether the authorities have the will, the knowledge and the resources to guarantee our safety.

1 / For What We Are About to Receive . . .

'My biggest problem is caterpillar shit,' said Peter Atkins, a forty-year-old fruit and vegetable grower from north Kent. 'It is green and slimy, but it washes off easily enough and doesn't hurt anyone. But if Marks and Spencer find any on my cauliflowers, they'll probably send back the whole consignment.'

Mr Atkins is hardly a fan of pesticides, although he is more knowledgeable than most about their use . . . and abuse. He belongs to Friends of the Earth and Greenpeace, credentials he prefers not to brandish among his farming friends. On the other hand, Peter Atkins also believes that organic farming is an irretrievable relic of a nostalgic past. And he, and his bank manager, know that if he wants to keep supplying the major supermarkets with his crops he must meet the stringent 'appearance' standards they require. We consumers, accustomed to the bright lights, polished floors and shrink-wrap packaging of supermarkets, are no longer prepared to accept green slime on our vegetables.

'I have got no choice but to use pesticides, a lot of them . . . more than I would wish . . . right up to harvest,' he says.

Conscientious farmers like Peter Atkins try to obey all the regulations and understand the potential dangers of the toxic chemicals they spray on their crops. Others, driven by commercial need, plain negligence or downright greed, don't.

Every time the authorities in the UK bother to look – and it isn't that often – residues of DDT and other banned or noxious chemicals turn up on the food crops we eat.

It would be wrong to assume that the alarm bells ring loudly in government circles, even though DDT and some of the more modern pesticides have been linked with cancer and birth defects in animals.

'In most circumstances, occasional exposure to higher-than-average levels of pesticide in a foodstuff has no public health significance.'

These were the reassuring words of the Ministry of Agriculture's most recent report on pesticide residues.* To be sure, most but not all of the pesticide residues found on foods in Britain appear in minute, almost undetectable quantities. But then, so do the carcinogens (cancer-causing agents) in individual cigarettes. And the Environmental Protection Agency in the United States, which regulates pesticide safety, says there is 'no safe level' of known or suspected carcinogens.

Pesticides are not the only powerful chemicals which farmers increasingly use to boost their yields and cut their costs. Growth promoting hormones, made in pharmaceutical factories and compressed into pellets, are still implanted into most of the beef cattle in Britain, despite the EEC decision to see them banned. In other animals like pigs and poultry the same artificial growth stimulation is achieved by adding constant doses of antibiotics to livestock feed.

We live in an age of chemical agriculture. We enjoy the benefits of lower costs. We are also the first generation of guinea-pigs who will determine whether these marvels of science and medicine are truly safe.

At least the authorities are beginning to require that the chemical additives which are intentionally put into our mouths by food processors are labelled on the can or packet – even if most of us would need a code-book and crash chemistry course to unravel the mysteries of the 'E' numbers and obscure compounds we find on a packet of biscuits or sausages.

Residues from pesticides, hormones and antibiotics – unseen and untasted – are not supposed to be in our food. So there is no warning about their use on the label – at least not yet. Even if no one intended that they slip unnoticed down our throats, they have not been put there by accident.

So let us make a closer inspection of that typical shopping trolley – filled with the food purchases for an average family of four – to see what chemical residues it might contain.

* Ministry of Agriculture, Fisheries and Food (MAFF), *Report of the Working Party on Pesticide Residues* (1977–81), p. 2

Milk

Across the social spectrum we spend, as a nation, more per week on liquid milk than on any other single food item. Milk is particularly susceptible to chemical residues. The dairy cow that produces the milk may well have grazed on pasture which has been sprayed with herbicides including 2,4,5-T, the now notorious weed-killer which is commonly contaminated with dioxin. The problem is that most pesticides are 'lipid-seeking' – that is, they cling to the fatty molecules known as lipids within our bodies. Milk is rich in lipids, which is why it is so fattening. Cows, in areas near towns, will also commonly graze on pastures which have been fertilized with 'sewage sludge' from waste disposal plants. The Ministry of Agriculture's own working party on residues admits that this sludge is often contaminated with persistent pesticides which pass down the food chain into the cows' body fat or milk as they graze. If this seems slightly fanciful, the same report admitted that the passage of pesticides down the food chain ends up, ultimately, in the breast milk of human mothers. Low levels of the persistent 'organochlorine' pesticides – DDT, Lindane and Dieldrin – were found in mothers who donated their milk to hospitals in 1979–80.* How these pesticides got there, no one is quite certain.

Dairy cattle only swallow pesticides indirectly, through the grass they graze on. Antibiotics are, potentially, a much bigger problem. This is because dairy farmers habitually use long needles to inject high doses of penicillin and other therapeutic antibiotics directly into the udders of their milk-producing animals. The antibiotics are commonly used to treat mastitis, an endemic udder infection. They are also routinely plunged into the udder, in what is known as 'dry cow therapy', to contain any residual infections during the periods when the cow is resting between lactations. Both treatments are open to abuse if milk is taken while high doses of the antibiotics are still being given. Fortunately, the Milk Marketing Board subjects milk to far more rigorous testing than is applied to any of our meat or fruit and vegetable crops. Unfortunately, the Milk Marketing Board set up the tests, not because it cared particularly about the consumer, but because it wanted to protect its own commercial investment in cheese production. Antibiotic residues in milk taken from a single herd can ruin the natural bacterial

* MAFF *Report of the Working Party on Pesticide Residues* (1977–81), pp. 22, 25

action of an entire milk consignment destined for cheese-making. Milk samples are supposed to be checked weekly by the MMB and heavy fines are imposed on farmers who let antibiotic-contaminated milk slip through to the cheese factories. Even so, the residue levels permitted by the MMB still exceeded the safe guidelines laid down by the World Health Organization of the United Nations until January 1986. What could these residues still do to the consumer? One researcher, now working as a drugs inspector on behalf of the government, wrote: 'Ingestion of milk (or other animal product) containing antibiotic residues could induce allergic reactions in the consumer.' And she warned, far more ominously: 'An extension of this problem is the possibility of sensitization, i.e., repeated exposure to small quantities of antibiotic leading to a severe reaction when given the same antibiotic therapeutically by a general practitioner.'*

The Milk Marketing Board apparently does no testing to discover excess hormone levels in milk. But cases have been reported where cull cows (ageing dairy cattle destined for the slaughter-house) have been implanted with growth-boosting hormones while they are still being milked on the dairy farm, and excessive levels of the hormones have been detected.

Beef

Before her milking days are over the dairy cow will give birth to perhaps a dozen calves. Some, if they are purebred and female, will enjoy a reasonable life in the dairy-shed like their mother, before they, too, are sent for slaughter. But most of the calves will be reared quickly to produce veal and beef, and they will meet a much quicker end.

These newborn calves, removed from their mothers within days, receive the full arsenal of chemicals. As soon as they are weaned they will be put on milk powders and then feed containing 'growth-promoting' levels of antibiotics. Young cattle lead a nomadic existence, picking up diseases as they pass from market to market. To combat or prevent infection, they are frequently given much higher doses of

* Wingfield, Joy, Pharmaceutical Society inspector, *Controls on the Use of Antibiotics in Animal Husbandry*, p. 35. See also Corry, J. et al., (MAFF officials) *Detection of Residues in Milk and Animal Tissues* (*Soc. for Applied Bacteriology Tech Series*, 1983)

'therapeutic' antibiotics on veterinary prescription. These drugs are often the same penicillins and carotenes which are the first line of defence for doctors in human treatment.

Distressed and weakened cattle are prone to the most virulent form of salmonella infection, called scours. An outbreak can kill an entire herd. The preferred, and still permitted, treatment in Britain is the antibiotic, chloramphenicol. But chloramphenicol is banned for food animal use in the United States for two apparently obvious reasons. First, the drug is the only effective treatment against typhoid in man. Research has shown that over-use of chloramphenicol in animal husbandry has led to increased drug resistance in the typhoid bacteria (a member of the salmonella family) which can be a killer in man. Fortunately, typhoid outbreaks in developed countries are now very rare. But the second reason for banning chloramphenicol is even clearer:

'This is because chloramphenicol can cause fatal aplastic anaemia and/or leukaemia in susceptible humans at any dose by any route of administration.' *

This emphatic warning came in June 1985 from Dr Lester Crawford, the director of veterinary medicines for the U.S. Food and Drug Administration. Yet it has not been banned for animal use in Britain.

Chloramphenicol may be a special case. Bacterial resistance to it from overuse does, quite directly, cause serious hazards in treating typhoid in man. And the slightest residue of the drug consumed by eating the meat or organs of an animal can, in Dr Crawford's view, cause life-threatening diseases like aplastic anaemia and leukaemia in susceptible consumers.

In the UK, because so few carcasses are tested, it is hard to say how much of the beef we eat is contaminated with traces of chloramphenicol. But veterinarians say that use of the drug (both legal and illicit) is widespread.

The other antibiotics used in animal husbandry certainly leave residues in beef, but direct problems, if they exist, are probably restricted to allergic reactions.

Of much greater concern, however, is the indirect problem of bacteria building up resistance to antibiotics prescribed for man because they are given to cattle and other animals excessively. Two of the

* Crawford, Dr Lester, *Speech to Food Editors Conference* (Dallas, Texas, 28 June 1985), p. 5

leading British scientists* in the field are convinced that these resistant strains, which are developed in calves and cattle, do cross-infect the human population. How great the danger is will be explored in subsequent chapters.

Cattle which survive the onslaughts of infection – and even those which thrive with relatively little recourse to antibiotics in their feed – still need to be fattened up before slaughter. Most of the beef which ends up on the British dinner table comes from cattle which have been implanted with manufactured growth hormone pellets supplied by the big drug companies. Farming is big business and growth hormones which have been approved as safe in Britain but banned in most other EEC countries can add nearly 100 pounds of carcass weight to a mature animal for very little cost. They also have the benefit of creating leaner meat for fat-conscious consumers. But controversy rages because residues left in the meat from one particular hormone (now banned worldwide) have caused sexual deformities in Italian children, and grave uncertainties remain about the ones still in use.

Poultry

A joint of beef, once the mainstay of Sunday lunch, is being replaced in many households by chicken – and with good reason. Chicken is excellent value. Its real price has fallen substantially over the last ten years, thanks largely to faster growing breeds and intensive livestock farming. Chicken also receives high praise from ecologists and health experts. Chickens convert grain into protein far faster than beef cattle and a bit faster than pigs, so they are a relatively efficient user of the world's scarce cereal resources. In addition, nutritionists praise the lean white flesh of poultry which contains far less saturated fat than red meats.

But this bounty is only bought at a price. Evolution never equipped the humble chicken to survive the cheek-by-beak coexistence of the indoor factory pen. One of the UK government's leading experts, Dr Bernard Rowe, estimates that eighty per cent of the chickens on supermarket shelves are contaminated by salmonella bacteria that must

* Professor Alan Linton at Bristol Medical School and Dr Bernard Rowe at the Public Health Laboratory Service

be killed by thorough cooking.* Man has made this hazard worse by breeding exclusively for fast growth to reduce 'days to market' while ignoring the durability of the birds he has created. Why should he worry when the problems of disease (at least in the chicken) can be contained by drugs? Chickens seem impervious to most strains of salmonella. What threatens them is the parasitic disease, coccidiosis. But drugs called coccidiostats effectively control the parasite which is endemic to all intensive poultry-rearing compounds. Fortunately, there appears to be little evidence that residues from coccidiostats cause any problem to the consumer. But factory farming is a cut-throat business, both for the farmer and the chicken, where profits depend crucially on quick weight gain. At least seventy per cent of the cost to the farmer comes from feeding the chicken. To cut his costs, the big poultry producer typically adds one of the 'growth-promoting' antibiotics to all his animals' feed from the day they peck their way out of the egg to the day they are slaughtered. These permitted antibiotic feed additives – like Flavomycin, Avotan and Eskalin – appear to work by killing off some of the natural flora of bacteria in the animal's gut.

This allows more of the feed to be converted directly into meat. The drugs also appear to have an 'anabolic' effect – that is, they increase the metabolic conversion of nitrogen compounds into protein.

These growth promoters, and other antibiotics used in therapy for sick animals, are no mere sideline to the pharmaceutical industry. At least half of all the antibiotics manufactured in the United States go into livestock, and the figure is probably only slightly lower in Britain.

Eggs

Factory egg farming, which confines laying birds to a cage for their whole working life, is unpleasant and, many believe, inhumane. In its most recent report the Farm Animal Welfare Council, which was set up by the UK government, roundly criticized the conditions which laying birds suffer.† Many consumers, particularly in the trendier regions of the south, are now paying a premium for guaranteed 'free-range' eggs. This effective protest may lighten the pockets and ease the

* Interviews with the author, July and December 1985

† Farm Animal Welfare Council, *Assessment of Egg Production Systems*, 16 September 1985

consciences of egg buyers. It may also help them eliminate some of the chemical additives which are added to factory feed to harden the shells, and colourings like tartrazine which improve the pallid colour of the yolks. But it will not significantly reduce the amount of drug residues consumers receive in their food.

Somewhat surprisingly, caged laying birds are subjected to relatively few antibiotics and no legally permitted hormones. The simple reason is that the drugs do not work and only cost the farmer money. Caged layers are not as prone to infection as birds for the table that need to put on muscle by running in colonies around their confined pens. The 'growth promoting' antibiotics are not much used either, since layers are typically full-grown and the boost they might get from the antibiotic does not seem to produce more or bigger eggs.

Even coccidiostats, the most common poultry drug, are usually not needed because the parasite is transmitted by the chicken's droppings, which conveniently fall through the wire mesh of an egg layer's cage. Table birds, by contrast, stampede over the droppings which collect with the litter and feathers in their pens. Still, in farming, almost everything has a commercial use. We shall see a bit later on how this pungent, but apparently protein-packed brew of droppings, litter and feathers is fed to cattle with the approval of both the U.S. and British governments.

Fish

The nice thing about fish, from the producer's point of view, is that they do not have to be fed – at least those caught from the sea or inland waters. This ought to help the consumer too since no one can intentionally meddle with their diet. But years of steady pollution have poured heavy metals like mercury, cadmium and lead as well as the persistent organochlorine pesticides like DDT into our lakes, streams and coastal waters. Inevitably they pass as residues through the food chain into the fish and shellfish we eat.

A growing problem, too, is the heavy use of antibiotics in factory fish farming. The rainbow trout, like the chicken, is an extremely efficient converter of feed to protein. But like the chicken, it is extremely susceptible to disease when crowded unnaturally into enclosed tanks. Unfortunately, little work outside the industry has been done on the use of drugs and their residual side-effects in fish farming.

Pork

Pigs have always had a bad reputation for hygiene and disease. This is a bit unfortunate since they are rather nice creatures – so nice, in fact, that they will, without much complaint, withstand being packed into sheds and being fed almost anything, including a constant regime of antibiotics to increase weight gain and prevent disease sweeping through the herd. Of all livestock animals, pigs are the biggest consumers of antibiotics.

'There are a very large number of pigs which are fed antibiotics right through to slaughter – and I am talking now about prescription-only drugs,' said a senior pig veterinarian from the north of England. 'There is no law against it so long as the vet signs the prescription. These prescription antibiotics are supposed to be withdrawn for a certain period before slaughter to reduce residues in the meat, but there is plenty of abuse.'

One of the pig farmer's favourite drugs has been Fortigro, a feed additive antibiotic. It has helped its manufacturer, the Pfizer pharmaceutical company of the United States, become the biggest supplier of animal drugs in the world. Doubts about its safety began, however, to nag officials in Whitehall. The Ministry of Agriculture, after allowing Fortigro's use for a number of years, has now proposed to ban it on the grounds that it is a 'genotoxic carcinogen', unsuitable for use in farm animals. The UK authorities believe that the biggest risk has been borne by feed mill and farm workers who were exposed to heavy doses of the drug. They do not believe that residues of the drug, eaten by the consumer in his favourite chops, joint or bacon, present any quantifiable hazard – even though there may be no minimum safe level of a proven carcinogen.

It may turn out that even trace doses of the drugs used in animal husbandry are harmful to the consumer – although at this stage such a dreadful prospect appears remote. It is more likely that any danger is dose-related.

Offal

If dose is important, then it is probably wise to consider how much kidney and liver you put in the shopping trolley. Once again, nutritionists advise us that offal is extremely good for us. It is low in fat, high in protein and what's more, it is relatively cheap. But even defenders of antibiotics and hormones admit that the organs of animals – particularly their livers and kidneys – will contain higher levels of the drugs than the muscle tissue we call meat. The reason for this is straightforward enough. Only a small proportion of the drugs fed orally pass from the gut wall through the bloodstream into muscle tissue. More collects in through the filtering organs of the body (the liver and kidneys) before it is excreted. Tests to prove the safety of drugs frequently show residue levels three times higher in kidney and liver samples than in meat. By this yardstick pork and calf offal should contain the highest levels of antibiotic residue, while the offal of beef might contain the highest residues of growth promoting hormones.

Lamb and mutton

On the meat front, at least, the good news comes at the end. Lamb and mutton are likely to contain the fewest chemical residues. Doctors concerned with allergies already know this and that is why they often allow lamb as the only meat in a patient's diet. Like poultry and pigs, lambs and sheep do not respond well to hormone growth promoters, so farmers find their use unprofitable, except in some cases to regulate the ovulation of ewes coming into season.

Growth-promoting antibiotics do not work either since most sheep in Britain are grazed openly in fairly remote regions where even herbicides to control weeds may not be used. Explaining the lack of drug use and abuse Mr Alan Davidson of the Pharmaceutical Society said: 'It is just too hard for the sheep farmer to try to catch the little buggers up on the hillside.'

Animals, at least, have their vocal defenders. The Vegetarian Society never tires in its efforts to wean us from meat – both for our own sakes and to avoid the death of livestock. The more strident animal-rights groups, preferring pillage to persuasion, raid factory farms and drug company laboratories. And far from the radical fringe, veterinarians are beginning to speak out forcefully about the abuses of livestock farming and chemical husbandry. The British Veterinary Association refuses to join extreme critics who want to see all hormones and antibiotics banned. Some of these drugs are, in present conditions, essential to the welfare of animals. But the BVA openly campaigns against the excessive use of antibiotics which breeds resistant strains of salmonella, infectious to man. It complains bitterly, although partially for reasons of self-interest, about the growing black market in prescription drugs used illicitly behind the farm-gate. And many veterinarians express grave doubts about the continued use of growth-promoting hormones in cattle.

Fruits and vegetables, however, have few committed supporters, with the exception of the campaigning environmental group, Friends of the Earth and the Soil Association. If crops are doused excessively with pesticides few, outside Whitehall, have a motive to tell us.

In addition, it is much harder to work out what has been sprayed on crops. More than 4,000 branded pesticides are licensed for use in Britain and a similar number are sold in the United States. In theory at least, the most hazardous are restricted to a very few conditions and crops. Others, deemed to be relatively safe, are allowed to be widely used.

Farmers are perfectly entitled, however, *to buy* whatever licensed pesticides they want without prescription or serious restriction. This may seem strange. After all, antibiotics designed to kill bacteria are controlled by the veterinarian's prescription – at least if they are used for therapy. But pesticides designed to kill insects, which are far closer to man on the evolutionary ladder, are sold without any real restraint.

There is no doubt that the food we buy in Britain is frequently contaminated with pesticide residues. In recent tests* the Association of Public Analysts found that more than one third of all fruits and vegetables analysed contained at least detectable amounts of pesticides. Some of the substances found, like DDT, were banned – that is, they were so hazardous that they had been banned completely, or their use was not permitted on the sensitive food crops upon which they were found.

* Nicolson, Ronald S., *Association of Public Analysts Surveys of Pesticide Residues in Food, 1983* (J. Assoc. Publ. Analysts, 1984), pp. 22, 51–57

Yet the government says we are not in danger because most of the residues were exceedingly low and 'in most circumstances, occasional exposure to higher-than-average levels of pesticide in a foodstuff has no public health significance'. Let's take a closer look at what the analysts found.

Salad vegetables: lettuce, tomatoes and cucumber

We used to wash lettuce to remove slugs, bugs and dirt. These days it would be wise to immerse it thoroughly in hopes of removing heavy doses of pesticide. Thirteen of thirty-two samples tested showed some contamination including 'reportable' levels of Lindane, a highly persistent organochlorine pesticide which has been identified in at least two independent studies as a carcinogen in laboratory animals.* Even more disturbing, in a separate study the analysts found that three lettuce samples were contaminated with high doses of DDT, the notorious insecticide which moved Rachel Carson to write her damning indictment *Silent Spring* in 1963. Governments around the world have banned DDT ever since, and although the UK authorities did not ban DDT completely until October 1984, its use on lettuce and other sensitive crops had been withdrawn long before the analysts took their samples. What is more, the levels of DDT found on two samples far exceeded the UK government's minimum reporting level and even exceeded the much higher 'maximum residue limits' set by the EEC but so far ignored in Britain.

Tomatoes and cucumbers escaped rather more lightly, but disturbing residues were still found. A third of the tomatoes were contaminated with pesticides, including a few with Lindane and one with Aldrin, another pesticide with a highly dubious safety record which has been banned for all agricultural uses in the United States, and is not permitted to be used on tomatoes in the UK. More than a third of the cucumbers sampled contained Lindane residues.

I have dwelt rather long on the salad crops because most of us eat

* Data linking commonly used pesticides with possible carcinogenic, teratogenic (causing birth-defects) and mutagenic disease has been compiled from recent academic studies by the London Food Commission.

these vegetables raw, and many of us seeing them so clean and pristine on supermarket shelves no longer even bother to rinse them before we eat them or serve them to our families.

The same caveat applies of course to fresh fruits.

Fresh fruits

Apples and pears receive a barrage of pesticides even before the first hint of the fruit is seen on the growing stem. One of Kent's leading fruit growers who supplies the major supermarkets sprays his orchards a dozen separate times before he finally picks his apples and pears. Those which are destined for cold store, to fetch a higher price in the coming months, are also 'drenched' in a fungicide, Benomyl, which coats each fruit. This miracle chemical prevents one rotting apple accidentally tossed in from ruining the whole bin by the time it is opened the following spring. All the chemicals he uses are government approved, proper doses and withdrawal times before harvest are observed, and he is so confident that he seldom washes the fruit his own family eats.

But tests by the analysts across the country revealed a different picture. Two apple samples from a batch of thirty were found to contain DDT above the government's reporting level, and altogether more than one in five apples had detectable levels of less unpleasant pesticides. Although no DDT was found on pears, more than half of the samples showed similar residues of other chemicals.

Soft fruits are less likely to be washed and are almost never peeled. Yet DDT and its close relative DDE were found on blackcurrants, strawberries and plums, and some cherries showed low residues of a banned form of Lindane.

Cooked vegetables

Cooking and peeling obviously reduces the potential hazards from pesticides. This is just as well, since the analysts found DDT at levels above the Government's reporting limits on roughly ten per cent of the brussels sprouts, cabbage and mushrooms they tested.

In total, pesticide residues of one kind or another were found on about a third of vegetables tested. How much washes off after cooking has never been clearly determined. This may seem odd. But all these chemicals have different properties. Some are stickier than others. After being sprayed, some break down quickly into harmless components. Others, like DDT, Lindane and Aldrin can linger for years in the food chain. Some disintegrate in the heat of boiling water. Others remain impervious. Besides, the government, advised by its experts, remains convinced that the safety margins built into most pesticides and other agrochemicals are so high that a few abuses here and there by farmers and the chemical and drug manufacturers do not really matter.

Take the example of fungicides sprayed on crops like potatoes, apples and pears, and oranges and lemons which are stored over long periods and require prevention from rotting before sale.

The government's own working party on pesticide residues in food quietly noted that some potatoes examined had levels of tecnazene, a sprout-suppressing agent, which were 218 times higher than the maximum residue level permitted by the EEC.* Washing and peeling were likely to take care of the problem, the experts said, apparently ignoring the fact that nutritionists are always advising us to eat baked potatoes in their skins to preserve the vitamins and minerals which the jackets contain!

Bread

The pesticide problem also confounds the sage advice of nutritionists to eat more wholemeal bread. Pesticides collect, naturally enough, in the outer, brown layer of the wheat kernel. Harsher milling to produce white bread presumably reduces them. And evidence presented to a British Crop Protection Conference in 1981 by two Ministry of Agriculture scientists showed that nearly fifty per cent of the pesticides on wholemeal wheat survived both milling and baking.†

It had been assumed, of course, that baking would have driven off or

* MAFF, *Working Party on Pesticide Residues* (1977–81), p. 11

† Residues of Organophosphorous Pesticides in Wholemeal Flour and Bread Produced From Treated Wheat, D. R. Wilkin and F. B. Fishwick, MAFF 1981

destroyed much more of the pesticides than it did. The work of the Ministry scientists threw open the whole question of how much residue from pesticides and the drug yield boosters actually remains unseen, untasted and largely untested in the whole range of processed and packaged foods which are beyond the scope of this book.

The Ministry scientists further admitted that the amount of pesticides they found in the wholemeal bread exceeded the maximum residue levels set by the World Health Organization and the Food and Agriculture Organization of the United Nations in three cases out of four. They nevertheless concluded that the residues found 'probably do not represent a toxicological hazard'. But they did recommend that the whole issue should be re-examined.

The increasing use of hormones, antibiotics and pesticides to boost yields in agriculture faces us with a growing problem whose more serious dimensions may not surface immediately. In contrast, the chemical and drug companies who make these products, the farmers who use them and the authorities who regulate them continue to insist that the consumer is in no real danger. To the contrary, they maintain that all of these growth promoters have been a positive boon to the consumer – bringing fresher, better looking and cheaper food to the table.

So how did these chemicals become a part of our daily diet? How were they discovered and developed and who are the scientists so convinced of their safety? A closer examination follows.

2 / The Guinea-Pig Generation

When the chairman of Ciba-Geigy, Dr Louis v. Planta, took the rostrum in the spring of 1984 to make his annual address to shareholders his words bristled with confidence. The men who run the giant Swiss drug and chemical combines like Ciba-Geigy seldom doubt the benefits their companies provide. And he knew that the unbridled assault he was about to unleash against 'the common enemy' would be greeted with uncritical acclaim by shareholders within the opulent auditorium.

The 'enemy' Dr v. Planta went on to attack was that motley assortment of environmentalists, consumer groups and churchmen who had the temerity to challenge the motives and good works of Ciba-Geigy.

'. . . If you think that our approach and behaviour receive the approbation of the public,' he told his audience, 'then you are mistaken. Quite the contrary is the case: we are subjected to constant criticism!'* And who might these critics be?

'I am referring to international consumer protection, environmental protection and Third World organizations, and also to the World Council of Churches . . . Their aim is to change society.'

Many people would like to change society. But Dr v. Planta revealed an insidious design that united all these protestors:

> Autocratically and often arrogantly they lay down the law on what is good or bad for other people, and answer to no man. Their declared objectives are often merely a front for aims that are fundamentally ideological. They attract numerous followers, including many well-meaning but misguided people . . . The tactics of the pressure groups are to use one-sided or even falsified presentations of concrete incidents or facts as a means of discrediting the multinational corporations and diminishing their credibility.

* *Address of the Chairman Dr Louis v. Planta, Chairman of the Board of Ciba-Geigy*, Ciba-Geigy Ltd. Annual Meeting, 10 May 1984

Well, that is one point of view. But there are still rather a lot of us, with no declared objectives and no particular desire to discredit capitalism, who would like to know why we and our children have, unwittingly, become the first generations to be fed a diet of chemicals concealed in the basic foods we eat.

The answer, at least for the growth-boosting range of pesticides and veterinary drugs, is that they were developed in an age when scientific achievement went largely unquestioned – and they have genuinely helped to feed the world by increasing yields and reducing disease.

Companies like Ciba-Geigy, which live in the closed and comfortable confines of the Swiss corporate state, would like to believe that only madmen and politically motivated extremists have grounds for questioning whether there is a hidden price to pay for all this bounty. When Ciba-Geigy wanted, for example, to test the unknown toxic properties of its new pesticide, Galecron, it chose to experiment on Egyptian children. In 1976 a group in Alexandria were paid a few pennies to stand almost naked in a field while a crop spraying airplane loaded with the chemical passed overhead to spray them. The company then took urine samples from the children to discover how much Galecron they excreted. One does not have to be a committed propagandist to question the behaviour of Ciba-Geigy in this human experiment. Nor does the company deny that it occurred, although it points out that it had permission from the local Egyptian authorities to hire and spray the children.*

Incidents like these are usually explained away as isolated errors of judgement. But it is also true that the big drug and chemical companies develop corporate climates where huge sums of money are gambled on new scientific discoveries. They do not like to see their fortunes dumped down the drain. The U.S. drug company, Pfizer, got extremely angry when the British government officially declared that one of the antibiotics it makes for inclusion in livestock feed, Fortigro, was a genotoxic carcinogen. In other words, the drug could leave potentially cancer-causing residues in meat eaten by consumers, and Britain wanted it banned. Pfizer, which fears the judgement may also hit its U.S. sales of the drug, accused the Ministry of Agriculture of ignoring mountains of scientific evidence in the drug's favour.

* The *Guardian*, 19 January 1983

This kind of debate could not have occurred twenty years ago. We had neither the analytical skills nor the desire as societies to challenge the bounty brought by the chemical revolution. After all, the real cost of producing a pound of chicken, a pint of milk, a dozen tomatoes and most other foodstuffs is far less than it was two generations ago. Some of the consumer's own cost benefit is undoubtedly eaten by the £12 billion worth of subsidies they must help pay into the Common Agricultural Policy of the EEC every year. Even so, people in Britain now spend only seventeen per cent of their national income on food. Before the last war the figure was forty per cent.

A substantial portion of this real saving can be traced directly to the pesticide and drug revolution in agriculture immediately after the war. Factory farming of poultry, which transformed chicken from a luxury to a cheap meal, could never have happened without routine feeding of antibiotics to prevent endemic disease running rife through the indoor pens. Hormone implants do enable cattle to put on weight cheaply and at an astonishing speed. And the pesticide makers claim that their chemicals have increased the yield of crops by up to thirty per cent.

EEC Meat Production (1982)*

	Extra tonneage from		*Total cost*	*Total net value*
	Hormones	*Antibiotics*	*of drugs*	*of extra meat*
Beef	170,000	86,400	£31 million	£160 million
Veal		50,400	£ 3 ..	£ 53 ..
Poultry		101,000	£ 3 ..	£ 65 ..
Pigs		400,000	£16 ..	£305 ..

* DSA, Bureau Européen d'Information pour le Développement de la Santé, Association International: *The Economic Contribution of Performance Promotants to Animal Production in the European Community*

Pesticides

The modern pesticide revolution only began after World War II. Farmers had, of course, experimented with simple pest killing compounds for centuries. The Sumerians discovered in 2,500 BC that sulphur could be used to control mites and other insects. An English essayist writing in 1825 urged farmers to pour common cooking salt on the crowns of weeds. And by the early years of the twentieth century three natural pesticides derived from plants (pyrethrum, nicotine and derris) had also come into fairly common use.*

But the vast arsenal of pesticides we have today owes more to man's fascination with gas warfare and his desire to exploit his newly-found oil wealth, than to any direct search for ways to improve food production.

The Russians formulated the first gas warfare compound, chloropicrin, in World War I. It was quite a success in combat . . . so long as the wind kept blowing in a favourable direction, toward the advancing German troops. When the wind reversed, however, the Russians got a whiff of their own weapon and collapsed with the same lung irritation and vomiting they had inflicted on the enemy. Before long the Germans developed their own phosgene version of the nerve gas weapon, and the 'organophosphorous' pesticides of today are direct descendants of this chemistry. Union Carbide was in the process of making organophosphorous pesticides at its Bhopal plant in India in December 1984 when a poisonous cloud of a chemical intermediate called methyl isocyanate (MIC) leaked out and immediately killed 2,500 people. Similar, if far less extreme, incidents occur all the time when farmers, using aircraft and tractors, allow their pesticide sprays to drift into neighbouring fields and villages. Friends of the Earth has compiled a dossier of injuries caused to people and animals after alleged spray drift incidents in Britain.†

It would be ludicrous to assume that the pesticide industry today is ignorant of, or unconcerned by, the inherent dangers of its products. But it is struck with the dilemma of trying to design molecules highly toxic to one form of life (the target species) yet harmless to man and the life forms he chooses to protect.

That impossible dream was apparently realized in 1939 when Paul Müller, a Swiss chemist working for the old Geigy half of Ciba-Geigy,

* Davis, Lee N., *The Corporate Alchemists* (Maurice Temple Smith, 1984), gives an excellent account of the development of pesticides and the entire chemicals industry. See also Sheail, John, *Pesticides and Nature Conservation* (Clarendon Press, Oxford, 1985)

† Rose, Chris, *Pesticides: The First Incidents Report* (Friends of the Earth, 1985)

hit upon the incredible insect-killing power of dichloro-diphenyl-trichloro-ethane, now known in every hamlet of the globe as DDT. Müller did not actually invent DDT. It was first synthesized in 1874 through a laborious process. But he learned how to make DDT simply and cheaply by incorporating lethal chlorine atoms into the convenient 'hydro carbon' molecules derived from crude oil. Müller's discovery helped waken the oil companies' interest in petrochemicals. He thought he had found the perfect pesticide: a cheap agent which attacked a broad range of insects yet appeared to be entirely harmless to man and the wildlife he wished to preserve. Better still, DDT kept on working long after it was sprayed. The older organophosphorous pesticides had the 'drawback' of not being persistent. We now know, of course, that DDT and its successor, 'organochlorine' compounds, are highly dangerous to man and higher life-forms because they are so persistent. They fail to break down in the environment, they get caught up in the food chain and they concentrate in the fatty tissue of man and other animals. But this is the clear vision of hindsight.

When DDT was discovered thousands of Allied troops were fighting their way through swamps only to die from malaria carried by mosquitoes and other tropical diseases. DDT was a life-saver which played an instrumental part, once the Americans and British learned to make it, in winning World War II.

Needless to say, the chemical companies that were dragooned into wartime production, were eager to exploit DDT's peacetime profit potential.

Some of the oil companies, like Shell, which had lent their expertise were also eager to get in on the new business. The world would need to be fed when the war ended and these companies correctly saw that the pesticide revolution was only just beginning. The opportunities seemed endless and the future problems of persistence and toxicity were still obscure. DDT was superb for killing the malaria mosquito and a host of other crop-eating insects. But the laboratories of the pesticide makers worked overtime to develop similar new compounds based on oil derivatives that would attack not just a broader range of insects, but also fungi and weeds. They succeeded.

Since the war pesticide manufacture, from virtually a standing start, has burgeoned into a global industry with sales of around £15 billion a year. How fast has it grown in Britain? That is a perplexing question.

Apparently, no one is sure. Since the government saw no real danger in pesticides in the early years, no one bothered to measure their use. It is believed, however, that UK pesticide sales rose from under £40 million in 1972 to about £300 million in 1985. Even after allowing for inflation, that means a four-fold increase in the amount of pesticides sprayed on the British countryside in just over a decade. More graphically, it also means that one billion gallons of pesticide spray (or 30,000 concentrated tonnes of active chemical ingredient) are used on UK crops and soils each year.

To say that the government saw no dangers in the early years is not to say that the dangers were invisible, or that good men of eloquence and power failed to report them. It is just that the ministries, in their wisdom, refused to listen and distorted what they heard.

In the early 1950s a special committee of eminent scientists under Lord Zuckerman was formed to look into this new science of pesticide manufacture. It was worried by what it saw. With considerable foresight Lord Zuckerman warned the government that tough new laws were needed to control the use and misuse of these powerful new chemicals. Public records recently disclosed show that the recommendations of Lord Zuckerman's committee were systematically altered or ignored by civil servants and his advice did not finally become law until the passage of the Food and Environment Protection Act in 1985. Even now the law only 'enables' ministers to require more stringent tests before new pesticides are approved for sale. It only 'enables' them to require retesting of many older pesticides which were approved before the potential toxic dangers of these compounds were understood. It only 'enables' them to set, if they see fit, maximum residue levels of pesticides in food. The new law *requires* them to do very little. And if they refuse to spend the extra money necessary to control the misuse of pesticides, then abuse of the burgeoning output will continue.

That abuse exists, which poses a threat both to human health and the planet's wider ecology, was first highlighted by Rachel Carson's book, *Silent Spring*, in 1963. The book was flawed: she made errors of fact and judgement. But her fundamental argument was sound and remains unchallenged to this day: pesticides are highly toxic chemicals which man spreads about the environment at his own peril. Ten years after Ms Carson published her book the U.S. government banned the use of DDT, the most persistent and insidious of all the pesticide compounds. Britain did not manage to ban DDT completely until 1984.

Today the uses of the other 'organochlorine' compounds like DDT are heavily restricted, and we have gone back, in a sense, to rely on the organophosphorous compounds which originated on the Russo-German battlefields of World War I. Today's compounds are less hazardous, but as Bhopal proved, they are hardly safe. We prefer them because their chemical bonds are less stable than the organochlorines and they break down into relatively inert components much more quickly.

This quick destruction undoubtedly lessens another problem caused by the persistent organochlorine compounds: insects are less able to develop resistance to their toxic properties. But the mechanism of death which the organophosphorous compounds employ while still active is quite horrific – as befits direct descendants of wartime nerve gases. In simple terms, they work by inhibiting the enzyme cholinesterase which is found in both insects and man. This enzyme allows muscles which have been told by the brain to contract to relax once their work is done. Without the enzyme the muscles of the body lock in permanent spasm. The anti-cholinesterase insecticides like Triazophos or Dichlorvos destroy the enzyme, paralyze the insect's muscles, and leave the victim to die.

It is not surprising, therefore, that insecticides (among all the pesticides) have been subjected to the greatest criticism since they are designed to kill other animal species, yet are sprayed directly on the food crops we eat. In temperate climates like Britain, most of the rest of Europe and the United States, the vast bulk of pesticides sold today are used to control weeds and fungi. A lot of the fungicides are sprayed directly on food crops – particularly in storage to control mould. Although herbicides are seldom sprayed directly on crops, some of them can be highly toxic. The herbicide, paraquat, is particularly lethal if swallowed in any significant quantity by man. ICI, the primary producer of paraquat, refused to sell supplies to the Florida state government in 1984 when the company discovered that the authorities intended to use the herbicide to destroy illicit marijuana fields in the state. ICI feared that smokers of the weed might inhale paraquat residues in minute quantities and did not want to be dragged through the American courts to face huge compensation claims if any of the 'pot smokers' fell ill. Another weed-killer which has been linked both with death and human deformity is the notorious herbicide, 2,4,5-T, which was used extensively as a jungle defoliant in the Vietnam War and is still permitted, for restricted use, in Britain today.

As the following table shows, herbicide sales in Britain still comfortably outstrip sales of fungicides and insecticides. But the underlying trend – which shows much faster growth of fungicides and insecticides – is disturbing. It is suspected that insects are learning genetically to beat even the less persistent organophosphorous sprays, and farmers are being forced to mount the 'chemical treadmill' – applying more and more insecticides, just to stay level. Fungicide sales have shown even faster growth (an eleven-fold increase in money terms). The reason is likely to have been Britain's entrance into the EEC and its contribution to the stored 'mountains' of food which the Community cannot sell. Fungicides are needed to attack highly dangerous moulds which grow on stored crops like cereals and nuts and produce extremely poisonous 'aflatoxins' – which can kill the consumer and have also been linked with cancer. Unfortunately, at least one of the fungicides which the chemical industry has produced, ethylene dibromide, has proved as dangerous as the fungi that it was designed to kill (see chapter seven).

Money Spent by UK Farmers on Pesticides (in millions of pounds)

	1973	*1976*	*1979*	*1983*
Herbicides	27.5	56.8	134.1	188.3
Fungicides	6.6	9.0	34.2	77.3
Insecticides	3.6	12.8	23.0	30.8

Defenders of the pesticide industry in Britain do, however, point with pride to its apparent safety record even during all the years that its activities were regulated only under the voluntary Pesticides Safety Precaution Scheme. On the available evidence, no one has ever died in Britain as a direct result of eating pesticide residues on food. And it has been more than a decade since an agricultural or factory worker was proved to have died from direct exposure to pesticides. However, in subsequent chapters we shall look critically at the refusal of the authorities to set up the kind of medical monitoring that could prove their claims that pesticides (and hormones and antibiotics) can be exonerated from causing undiagnosed sub-acute illness, chronic disease and even death.

In rebuttal, both governments and the industry point to the arduous and expensive testing procedures before a pesticide is unleashed on the market. A corporation can spend up to £20 million and ten years from the time of patent trying to get a pesticide past all the required safety hurdles. The time, scientific effort and sheer cash spent are our best guarantee of safety, they claim.

But the critics look down the other end of the telescope. What they see is tonnes of intentionally toxic chemicals, which have never been proved to be safe to man, being spewed out and sprayed on crops and around the countryside with no effective hindrance. After all, drugs like antibiotics, which kill single cell organisms, are first tested directly on volunteer patients and then strictly controlled by doctors' prescriptions before they are used at all. Contrast these controls with those governing insecticides – noxious chemicals which have never been tested in man, which are designed to kill evolutionary cousins much closer to man than bacteria are, and which are sold in any bulk required to any farmer with ready cash and a lorry waiting to cart them away.

In addition, the safety hurdles designed to keep dangerous pesticides off the market are not without their flaws, and they can be side-stepped. Governments do not test pesticides. They rely almost exclusively on the safety data supplied by the chemical companies. The Environmental Protection Agency in the United States has few laboratories of its own and the Advisory Committee on Pesticides in Britain has none.

A scandal which still reverberates began to erupt in the late 1970s when a major independent laboratory, International Bio Test of Illinois, which conducted experiments with animals to prove pesticide safety, was found to have rigged the data of the chemical and drug companies for which it did contract work. A World in Action television programme on ITV estimated that International Bio Test was responsible for about one third of all the animal lab studies done on pesticides and drugs in the world.*

Large numbers of pesticides and drugs were approved as safe – partially on the basis of IBT studies – by the drug and pesticide authorities in Britain and the United States. Three of the senior ex-

* World in Action, 'Tried, Untested', transmitted 17 December 1984

ecutives of the company were subsequently given jail sentences after the U.S. courts heard that many of the animal tests were entirely bogus. Animals which died in their cages, while being given the drugs or pesticides, were simply thrown away and fresh animals were substituted just before the end of the trials to prove none had been harmed. In other cases the data was simply invented.

When the fraud was finally uncovered by an alert official at the U.S. Food and Drug Administration the authorities moved swiftly to order new tests on all the suspect drugs and pesticides. But none were removed from the market – either in the U.S. or in Britain – while the review, which took years, was being conducted. In Britain alone, thirty-three pesticides were implicated in the IBT scandal – including paraquat and Roundup, the biggest selling herbicide in the world, made by the U.S. firm, Monsanto.

No one, except ministry officials, scientists and the manufacturers was allowed to see the flaws in the original data and judge whether the new tests genuinely guaranteed the safety of the products. The pesticide and drug companies, which were protecting sales worth hundreds of millions of pounds, lobbied hard and effectively to keep their wares on sale and the review secret.

The International Bio Test scandal has contributed to a public fear of pesticides and growing scepticism about the real safety of drug and chemical testing. In this atmosphere of suspicion pressure groups in Britain, led by Friends of the Earth and the Freedom of Information Campaign, struck back. They managed to extract some apparent guarantees from the UK government during passage of the Food and Environment Protection Act 1985 that more safety data would be publicly available in future before new pesticides came on the market. But these assurances, which fall well short of a firm statute, have yet to be tested. And sceptics believe that the manufacturers, who worry about their commercial secrets, have also been given assurances that the crucial raw data of their experiments will never be revealed.

Growth hormones

Whatever else may be said about the safety of pesticides, they do have at least one comparative point in their favour: the residues they leave tend to remain on the surface of the food we eat. Those cleared for use on crops are designed not to penetrate deeply into the plant, and washing helps to reduce residues.

But this is not true of the growth-promoting drugs which are implanted into or fed directly to livestock. These hormones and antibiotics penetrate deeply. They flow through the animal's bloodstream, into its organs and end up deeply embedded as residues in the meat we eat and occasionally even the milk we drink. There is no debate about this. The drug residues are there. The only question is whether they are harmful in the amounts in which we swallow them.

The first growth hormone created in a laboratory was synthesized by British scientists back in 1938. Its full name is diethylstilboestrol – a mouthful which is commonly shortened to the initials, DES, or referred to generically as stilbenes. Stilbene is a name to remember because the human deaths and deformities created by these hormones have blackened the reputation of all the other growth hormones still in wide use. These tragedies began to be known in the middle of the 1970s. Before that, growth hormones, like pesticides, had been an unchallenged boon. And like pesticides, their economic importance in boosting food yields lay dormant until after World War II.

The bonanza could not begin until scientists learned to make replicas of the natural hormones in large batches in pharmaceutical factories. Once the secret of production was out and bulk sales could be achieved, farmers in Europe and the United States were encouraged to grab all their livestock and try the new hormones out. Farmers, being canny, worked out fairly quickly when they were throwing their money away. Hormones to promote growth have been tried on sheep and pigs, but the weight gain seldom justified the money spent. Why the drugs did not work is still not clear, and research continues. Imperial Chemical Industries did develop a product called Aimax which seemed successfully to stabilize the breeding cycles of sows. But laboratory tests indicated that an alarming number of piglets were born with congenital defects and the company withdrew it from the market in 1971.

Poultry, however, looked a splendid target for hormones. Intensive farming methods were rapidly bringing down costs and making chicken ever more popular. Indoor rearing also meant that the birds could be captured and handled easily. Suddenly consumers were being told that capons – bigger than an ordinary chicken but small enough for a family meal – were the perfect answer for Sunday lunch. Capons are male birds whose testes have been left intact. But to speed their weight gain a pellet of synthetic female hormone, hexoestrol, is inserted under the skin in their necks. For at least two decades many of us dined on this strange hermaphrodite and thought little about it. But hexoestrol is a member of the stilbenes family, and growing concern about the hormone's safety led poultry producers to look for alternatives before the substance was officially banned in 1982.

The answer they came up with was improved breeding techniques and the use of growth-promoting antibiotics. The average table bird, weighing three and a half pounds, now lives for only forty-nine days before it is big enough to slaughter. At this tender age it is effectively sexless and hormones would not work anyway.

Today in Britain and the United States legal hormone use is restricted primarily to beef and veal production. In Britain more than half of the beef on our tables has been implanted at least once in its life with growth hormones. In the United States the figure is said to be ninety per cent.

The primary reason why hormones are used and still so zealously defended in Britain (and Ireland) is that it is standard practice to castrate bulls. Elsewhere in Europe, with the exception of France, castration is almost never practised. So if the EEC ban does come into force in 1988/89, most continental farmers will feel little effect. Castration sounds unpleasant and cruel. But it does have genuine advantages which even environmentalists and animal welfare groups cannot ignore. On the continent bulls must be locked up to be fattened on grain for slaughter lest they run amok among grazing herds of contented heifers. The British bull may have become a eunuch, but at least it enjoys open pastures and the freedom to roam. By eating more grass it needs less grain which ought, on economic and moral grounds, to be fed to people. But the castrated steer does not put on weight very well – unless it is given hormone replacement therapy. Studies also appear to show that the castrated steer which is given hormone implants produces leaner meat with lower fat content than a normal bull – just what the nutritionists, concerned about saturated animal fats in our diet, have ordered. And defenders of hormones, which include most leading government scientists, also claim that the implanted steer will have less hormone residues in its meat than an intact bull.

Yet controversy about the safety of hormones continues to rage. The whole issue is muddled by the simple fact that all animals – be they livestock or human, infant or old – are constantly producing their own mixtures of male and female hormones in varying quantities. What harm is there in adding a bit more to the animals we want to fatten quickly for slaughter, if the extra amounts we might consume are tiny in relation to the amounts of hormones we already produce? Defenders of hormones claim that a prepubescent boy produces 1,000 times more male hormones in a day than he would consume from a beefsteak taken from a properly implanted animal.

Nevertheless, only five western countries – the United States, Canada, Britain, Ireland and France – permit all five of the hormones now on the international market to be used within their borders. They are all banned in Germany (which has the strictest food regulations in Europe), Holland, Belgium, Italy and other meat-producing countries. These countries argue that all hormones – whether natural or manufactured – are highly potent, very complex and little understood substances whose cancer inducing potential has not been fully explored.

The debate has been made even more complicated by the introduction of the two so-called synthetic hormones: trenbolone acetate (made by Hoechst under the trade name Finaplix) and zeranol (made by the International Minerals and Chemical Corporation under the trade name Ralgro). In fact, all the growth hormones in use are 'synthetic' in the sense that they are manufactured in drug plants. Three of them – testosterone, oestradiol and progesterone – are termed 'natural' because they are replicas of the naturally occurring male and female hormones in mammals. Finaplix and Ralgro are entirely synthetic, laboratory inventions. They are not naturally made by any animal.

So, the full line up of growth-promoting hormones now in use looks like this – see table on page 41.

Even this list simplifies the issue because cattle and veal calves are frequently given a combination of different hormones depending upon their age, sex and disposition. Castrated steers appear to thrive, curiously enough, on a pure boost of female oestrogen, but they are more commonly given a mixture of male and female hormones. Sim-

Compound	*Type*	*Origin*	*Name*
Androgens	Male Sex Hormones	natural synthetic	testosterone trenbolone (Finaplix)
Oestrogens	Female Sex Hormones	natural synthetic	oestradiol – 17B zeranol (Ralgro)
Progestagens	Female Sex Hormones	natural	progesterone

(Source: G. E. Lamming, Journal of the Royal Society of Health 1:1983, p. 9)

ilarly, heifers and cull cows are often implanted with Hoechst's synthetic male hormone, Finaplix. And the makers of Ralgro claim it works wonders on cattle of either sex. Female veal calves get a blend of male and female hormones, but their brothers, curiously again, get two kinds of female hormone – oestrogen and progesterone, even though the latter naturally functions to prepare the womb for pregnancy.

It may seem strange that it is necessary to manipulate an animal's sexual system to get it to grow faster. This is because all the hormones work in two different ways: they stimulate sexuality, but they also have an anabolic effect – hence they are also known as anabolic steroids. This point has not escaped weightlifters and other athletes who are reputed to use illegally an injectable version of Finaplix to put on muscle.

The natural hormones have long been known to cause deformities of the sexual organs and extreme surges of arousal in some implanted animals. Farmers refer to heifers in this state as 'bulling'. It means they try to mount the other females in the herd.

Hoechst and IMC, the companies which make synthetic Finaplix and Ralgro, claim that their products are superior because their effect is more anabolic (bringing about weight gain) and less sexual (causing awkward disturbances). Certainly some independent scientists would agree with them. But buried rather deeply in Hoechst's own promotional literature for Finaplix the company warns that the hormone should not be used on animals intended for breeding and it admits to the following 'contra-indications'.

. . . When trenbolone acetate (Finaplix) was given to female Friesian calves in an attempt to accelerate growth and development, reproductive performance was impaired. The onset of puberty was delayed, clitoris size was greatly increased, whilst vulval length was reduced . . . A reduction in functional mammary tissue has also been observed.*

Studies examined by the European Commission appear to indicate that Finaplix has low 'oral toxicity'. In plain language this means that most of the residues we eat in meat should be broken down and rendered harmless by our digestive system. Should be. But the same examination failed to discover a minimum dose at which no 'hormonal activity' in test animals could be detected.† In other words, it is unlikely that eating Finaplix residues in meat will cause the gross deformities suffered by the Friesian calves in Hoechst's own experiments, but there remains no absolute proof that the drug, when eaten, has no hormonal or other adverse effects in humans.

The same examination of Ralgro was even more inconclusive and alarming. Dogs fed Ralgro showed 'evidence of gross hormonal effects at all levels tested. A no-hormonal effect level was not determined.' ‡ As a result, the manufacturer was told to conduct new long-term studies but these will not be completed until 1987 at the earliest. Yet both Finaplix and Ralgro will remain on sale at least until 1989 in the UK, and Ralgro was only brought under veterinary prescription in Britain in 1985.

Much of the justifiable suspicion about the safety of these new synthetic hormones stems back to the medical use of the first synthetic hormone, DES – stilbenes – between 1940 and 1970 in both Britain and the United States. Pregnant women were injected with the hormone by their doctors in the mistaken belief that it could prevent unwanted miscarriage. The result was 'a medical disaster of the magnitude of the Thalidomide tragedy', according to a report which appeared in a recent British medical journal.§ Just as in the Thalidomide scandal, most mothers and their babies escaped without damage. But at least 429 girls born to these women contracted vaginal

* Finaplix, promotional literature, Roussel/Hoechst, p. 16

† European Commission, *Proposal for a Council Directive . . . concerning the prohibition of certain substances having a hormonal action . . .* (Com (84) 295 final, 12 June 1984) pp. 7–10

‡ European Commission, *Proposal for a Council Directive . . . concerning the prohibition of certain substances having a hormonal action . . .* (Com (84) 295 final, 12 June 1984) pp. 10–12

§ Emens, Michael, British Journal of Hospital Medicine, Birmingham and Midlands Hospital for Women (January 1984), pp. 42–46 cancer when they were in their teens or early twenties. One little girl developed a tumour when she was only seven, and at least seventy-nine young women have died. Campaigners seeking compensation for the families of victims say the true death figures are much higher and there is also considerable evidence that many mothers themselves developed premature cancers although this is harder to prove. A high proportion of the cancers and deaths occurred in the United States where doctors used stilbenes in pregnancy much more liberally than in Britain.

As we will see in the next chapter, the human suffering and deformity caused by stilbenes has not been limited to pregnant women who were injected with the drug and their daughters. Although the medical disaster was finally proven in 1971 and direct injection of the hormone ceased, the authorities remained convinced that stilbenes could still safely be used as growth-promoting implants in animals. As a result, young children who merely ate stilbene residues in baby food continued to suffer sexual deformities in the decade that followed.

Yet the authorities in Britain and the United States seem to believe that these tragedies cannot happen again. Hormones still in use are safe now because our knowledge is so much greater. The authorities almost behave as if the scientific errors which led to the stilbenes tragedies occurred in the dark ages of alchemy rather than just a few years ago.

Their confidence stems from the belief that the tiny residues of today's 'low oral activity' hormones represent only a small fraction of the total amount of sex hormones our bodies are subjected to every day. Meat from pregnant heifers (which are commonly and legally slaughtered) contains eight times more oestrogen than a steer which has been properly implanted with Ralgro. And four ounces (100g) of wheatgerm surprisingly contains thirteen times more oestrogen than a six ounce (160g) steak from the same implanted steer.*

In reply, consumer groups in Europe and the United States say that the possibility that even the tiniest amounts of the synthetic hormones (like stilbenes) can cause cancer has not been ruled out. Furthermore, the authorities always hedge their assurances with the proviso that the hormone implants must be used strictly according to their instructions. But in reality there is ample evidence that hormone implants are frequently over-used, withdrawal times are ignored and the pellets are

* Lamming, Prof G. E., *Growth-Promoting Hormones, Journal of the Royal Society of Health* (1:1983), pp. 8–11

embedded in edible parts of the animal. Finally, the critics want to know why the hazards of hormone implants are tolerated when mountains of surplus beef are stored at huge cost to the taxpayer in Europe and when surplus production has ruined the livelihoods of so many cattle ranchers in the United States.

Antibiotics

Antibiotics, even more than pesticides, were seen as a wonder cure for all problems in the 1950s. Pneumonia, dysentery and all manner of rampant infections could miraculously be defeated and lives could be saved once antibiotics came into common use in the 1940s. The problem of allergy to these powerful drugs was little understood and the graver problem of creating resistant strains of bacteria lay in the future.

Little wonder then that the Americans, with their characteristic commercial zeal, even began adding antibiotics to chewing gum and toothpaste as soon as supplies became available.* A daily dose was good for you!

The authorities in Britain were more cautious, for a while. But by 1953 farmers were allowed to include penicillin and tetracycline into some animal feedstuffs, even though the drugs were and have remained on prescription in human medicine. Initially these drugs were used either to cure or prevent disease – a growing problem as intensive livestock methods were developed to feed an impatient nation tired of the austerity of rationing. But it did not take farmers long to notice that the antibiotics, when used regularly, actually had the extra bonus of causing apparently healthy animals to put on weight more rapidly while eating less food.

To this day it is not entirely understood why antibiotics have a 'growth-promoting' action. The common explanation is that they kill off some of the natural and harmless bacteria in the gut and so allow more of the feed to be absorbed and converted into meat. Farmers cared little about the science, more about their increased profits, and use of antibiotics as growth promoters grew rapidly despite the increasing concern of many scientists.

* Howie, Sir James, *Ten Years on from Swann* (The Association of Veterinarians in Industry, London, 1981)

In the 1960s Britain was hit by successive outbreaks of salmonella among young cattle. In several epidemics the bacteria proved themselves quite resistant to antibiotic treatment. Cattle in their thousands died and a worried Labour government set up an expert committee under the chairmanship of Professor M. M. Swann in 1968.

The report of the Swann Committee in the following year became the bench-mark throughout the world for the proper control of antibiotics in animal husbandry. Its findings and conclusions were so clear and succinct that it is best to quote directly from the report:

Some enteric organisms, particularly in the salmonella group, are able to cause disease in man and also in some species of farm livestock. A notable example is *Salmonella typhimurium*. It is disturbing to note that the tendency for this organism to give rise to generalized infection in man has increased, for such cases require antibiotic treatment. If however, the strain of *Salmonella typhimurium* of animal origin shows multiple resistance to antibiotics, treatment by this means may not be possible and in the absence of other suitable treatment the life of the patient may be endangered.*

The Swann Committee went on to warn that salmonella bacteria were not the only culprits. All kinds of organisms, many of them quite harmless, could easily acquire resistance in the guts of animals fed on a steady diet of potent antibiotics. These common bacteria, like *E. coli*, constantly invade man when he eats meat. Once inside the human gut they had the ability to transfer their learned resistance to antibiotics both to man's normal bacteria and directly or indirectly to dangerous organisms, like typhoid bacillus, which he may have contracted. Gravely, the Swann Report continued, 'Such a chance meeting between resistant organisms and highly dangerous (pathogenic) ones could give rise to a potentially explosive situation.'†

To prevent a deadly outbreak of disease which doctors might be powerless to cure, the Swann Committee attempted to divide antibiotics for animals into two inviolate categories:

THERAPEUTIC antibiotics – those which are also used as front line drugs in human medicine like penicillin, tetracycline and chloramphenicol. Farmers would only be allowed to use these on a veterinarian's prescription to control a known outbreak of disease.

* *Joint Committee on the Use of Antibiotics in Animal Husbandry and Veterinary Medicine*, Report 1969 (the Swann Report), p. 60

† The Swann Report, p. 60.

FEED antibiotics – those which have been discarded for use in human medicine. Farmers could buy these 'growth promoters' without prescription. But the doses given daily in feed should be low enough to prevent bacteria from acquiring resistance.

The main recommendations of the Swann Committee were incorporated into the Medicine Act and became law in Britain under the Heath government in 1970. But in the United States, which prides itself on the effective regulation of health hazards, the crucial distinction between human-prescription-therapeutic antibiotics and the freely available growth promoters has been rejected by Congress – despite pleas from the U.S. Food and Drug Administration. Penicillins and tetracyclines are still poured without restriction into the feed of farm animals.

The trouble in Britain is that outbreaks of resistant salmonella kept occurring even after Swann was on the statute book. Had the wise professor and his colleagues been looking down the wrong end of the microscope? No: but they and successive governments failed to realize how simple it would be for both farmers and drug companies to circumvent the spirit of the law . . . with a little help from veterinarians.

There is no doubt that even healthy pigs, poultry and cattle grow faster on a steady diet of antibiotics. But the permitted growth promoters, like avoparcin, lincomycin and virginiamycin, are relatively weak. This is because they are orphans of the pharmaceutical industry antibiotics researched for the more lucrative side of human medicine which were cast aside when they proved ineffective in curing disease. The temptation to use the restricted human antibiotics, which offered the prospect in higher doses of even bigger profits was, and remains, great. And so, more by trial and error than design, was born the vast grey area of 'preventative' animal medicine.

If a farmer could convince himself and his veterinarian that his animals had been exposed to a disease threat and *might* become sick, then he could persuade the vet to write out a blanket prescription for the restricted therapeutic antibiotics . . . just in case. Here is a loophole which the Swann Committee failed adequately to consider, and one which remains flagrantly open today.

The greatest abuse appears to occur in cattle rearing and in pig farming. Cattle, because they are removed as only day-old calves from their mothers and forced as tiny animals to travel the country in enclosed lorries, are most susceptible to disease. Farmers who sent sick young cattle to market risked not getting paid. To protect their profits some began sending young calves on long and exhausting journeys with supplies of nitrofuran (a powerful antibiotic) tied around their necks with instructions to use it if any evidence of scours (violent diarrhoea) appeared. Unfortunately, young cattle are the carriers of those strains of salmonella which are most virulent, most resistant and most able to cross-infect man.

In 1964 there were just 4,500 cases of salmonella food poisoning reported to the Public Health Service Laboratory. By 1983 the figure had soared to 17,000. And, according to Dr Bernard Rowe at the PHSL, for every one notified case probably a hundred people suffer acute stomach upsets caused by salmonella without reporting it to the authorities.*

Pigs, despite their unfair reputation for poor hygiene, are actually very clean creatures. Alone among livestock they will search out and agree among themselves a separate place to defecate if given the opportunity and space. But pigs, herded together, are highly susceptible to disease. And in the squeezed confines of the modern industry, where profits are poor, they are not always given the chance to be so tidy. As a result, more than half of all the antibiotics administered to farm animals are given to pigs.

At the end of the day veterinarians are responsible for this misuse of antibiotics. They are only supposed to sign prescriptions for animals 'in their care'. But as we shall see in subsequent chapters veterinarians are put under enormous pressure – by farmers, feed manufacturers and the drug companies themselves – to sign prescriptions for animals they have never even seen.

The greatest abuse is in the administration of the antibiotic, chloramphenicol. Britain's leading expert on drug residues in food, Dr Ray Heitzman, has described chloramphenicol as 'the least prescribed, most used drug in modern (animal) medicine'.†

Farmers like chloramphenicol for the same reason that they liked the now-banned stilbene hormones: it is cheap, easy to administer and it works. And if vets get squeamish about prescribing, most farmers know how to get drugs on the flourishing black market, even if the

* Interview with the author, 8 August 1985

† *Farming News*, 22 March 1985

more scrupulous do not indulge in the illicit trade themselves.

Thus far we have looked at the problems caused by bacterial resistance to antibiotics. But are antibiotics themselves, as residues in meat and milk, also a danger? The answer is no one is entirely sure. Certainly even trace amounts of chloramphenicol appear to be a potential health hazard. And milk, taken from a cow which has just received a shot of antibiotics certainly contains enough drug residues to destroy the friendly bacteria used in cheese-making. But the authorities in Britain and the United States insist that the residues appear in such small quantities that they should present no danger to the consumer.

However, researchers for the Ministry of Agriculture recently challenged this conventional wisdom. They found that antibiotic residues 'could be responsible for triggering an (allergic) response in previously sensitized individuals'. And they added rather ominously: 'Allergic responses may be more common than is generally realized, due to failure to identify the cause, or attributing allergic reactions to food itself.'*

The safety of animal medicines is reviewed in Britain by the Veterinary Products Committee. It is designed to work exactly like the Committee on the Safety of Medicines which approves the safety and usefulness of human drugs for the Department of Health. But there is an essential and quite crucial difference. The Veterinary Products Committee answers to its superiors within the Ministry of Agriculture. The trouble is that the Ministry of Agriculture's prime role, as defined by law, is not to worry about the consumer but to sponsor Britain's farmers and to promote the rapid expansion of domestically-grown food. In the United States control of veterinary drugs is taken out of the hands of the U.S. Department of Agriculture, and given to the Food and Drug Administration.

In Britain, the Ministry of Agriculture has never shown itself very eager to pry into farmers' business. A host of voluntary codes and regulations are supposed to ensure that agricultural drugs and chemicals are used properly so that residues in our food are kept to a minimum. But what actually happens remains largely guesswork. The regulations about how to use these potent chemicals are complex. Most

* Corry, Janet E. L., et al., MAFF in *Society for Applied Bacteriology Technical Series* (1983)

farmers are very busy, and more than a few are complacent. Evidence that farmers frequently overuse drugs and pesticides, or fail to observe withdrawal times, is widespread.

But the question still remains: Can this misuse of chemicals in our food harm us? The best way to answer may be to step back for a moment and look at the tragedies which have occurred with drugs like thalidomide and Opren in human medicine. Both of these drugs went through extensive clinical trials in volunteer patients before they were ever allowed to go on the market. Yet one still proved to be a deformer of babies in the womb and the other a killer among the elderly. What clinical trials involving human volunteers have ever been conducted on pesticides and the synthetic hormones now in constant use? The answer, of course, is none. We have no direct evidence, carefully monitored, of what these potent compounds can do to man.

Drugs like thalidomide and Opren were also given in high doses, for a short period of time, to a known group of patients and doctors were alerted to be aware of their side-effects. We all consume chemicals used in agriculture at a low dose over our lifetimes, without prescription and without the particular knowledge of our doctors.

Dr Bill Inman at the University of Southampton is Britain's leading expert on human drug surveillance. He knows how hard it is to spot a rogue drug going wrong without bodies in the morgue as epitaphs. So he recognizes the needle-in-a-haystack task of trying to spot illness or injury caused by low levels of chemical residues in food. 'The chances of identifying a chemical residue as the proven cause of an adverse reaction are very remote – even though the chemical could occasionally be causing very real harm in some people.'*

These dangers are greatly increased when farmers, whether through greed or sheer pressure to meet production schedules, ignore the rules and overdose our food with drugs and pesticides. Then the tragedies are not mere conjecture. They are real.

* Interview with the author, August 1985

3 / The Human Tragedies

Milan, Italy, 1980

It began at first as an embarrassing secret – the kind of thing that parents the world over, fearing the worst, choose to ignore. Infant boys in Milan were growing breasts and baby girls were developing the protruding labia of maturing adolescents. Isolated and frightened mothers at first blamed their imaginations.

But the epidemic which turned toddlers into sexual freaks soon spread to Tuscany and Bergamo. By the summer the news was out. A Milan paediatrician reported to the authorities the case of a baby boy less than a year old 'with real breasts and dark nipples', and baby girls of similar age with 'very well-developed sexual organs'.*

After some initial confusion the Italian health inspectors discovered the common link between all the children: each had been fed jars of baby food containing veal which had been heavily contaminated with the female growth hormone, stilbenes. The drug was illegal. The Italians had decided to ban stilbenes shortly after evidence of vaginal cancer in the daughters of women treated with the drug began to emerge from the United States in the early 1970s. On 3 September the Italian minister for health ordered an immediate ban on twenty-two popular brands of veal based baby food.

Since stilbenes were supposedly banned in Italy and the existence of a black market was officially unrecognized, the authorities began to search for a suitable scapegoat outside their borders. They hit upon France, which supplied a third of the Italian market with veal. But the French had banned stilbenes too. Indeed, almost alone among European countries, Britain had allowed the continued use of stilbenes even

* *France-Soir*, 6 September 1980

though medical journals in the UK had, from the outset, sounded the alarm about the hormone's dangers.*

The British can easily be accused of official complacency. But at least they were not guilty of hypocrisy. Elsewhere in Europe it was an open secret that livestock farmers were having no trouble getting hold of illicit supplies of stilbenes to use liberally in fattening their calves and cattle. Even *France-Soir*, the newspaper which gallantly defended French honour in the scandal, had to admit that stilbene use in France continued because 'certain breeders cannot resist the temptation of watching their animals gain weight rapidly'.

Stilbenes had always been the farmers' favourite. Unlike the other hormones which are difficult to make, stilbenes can be concocted in any backstreet laboratory by anyone with a basic knowledge of chemistry. So they were, and remain, cheap to buy. They have other attractions. Stilbenes offer unusually quick weight gain and they can be injected directly into the muscle tissue of livestock, where they remain hidden from visual inspection.

In response to the Italian scandal the European-wide consumers' union, BEUC, which is partly funded by the Consumers' Association in Britain, launched an immediate veal boycott throughout the EEC. It also began to investigate how the stilbenes users had got round livestock inspections. Farmers had been in the habit of injecting the hormone directly into the necks of calves.

> But as the meat inspectors took their samples mainly from this spot, in order to obtain positive control results, the stock farmers started injecting cattle almost anywhere, in the most unlikely places (base of the tail and even testicles). Consequently, the least valuable parts of the animal (neck, base of the tail, breast) which are used in the manufacture of baby-foods, proved to be the richest in hormones.†

The hazard was compounded because the injected stilbenes are not rapidly absorbed and dispersed into surrounding tissue. They stay in a pocket within the tissue site and act as a natural slow release reservoir into the bloodstream. This explains why most of the children who ate the twenty-two brands of baby food on sale in Italy were unharmed.

* 'Stilboestral and Cancer', *British Medical Journal*, 11 September 1971

† *Black File on Hormones and Antibiotics*, Bureau Européen des Unions de Consommateurs, 1981, p. 13

Only those babies unfortunate enough to have been fed from a jar whose meat had come from the injection site, or very near it, received the massive doses of the hormone which caused their grotesque sexual deformities. As far as can be established, all the babies ultimately returned to normal when their diets were changed.

Puerto Rico

Young children in Puerto Rico have been far less fortunate. Two years after the Italian scandal surfaced, strange and disturbing letters were published in the *Lancet*, the leading British medical journal, from a team of paediatric endocrinologists from De Diego Hospital in San Juan. They reported hundreds of similar cases of thelarche – the medical term for abnormal premature sexual development in young children. The suspected cause was again stilbenes. But the alleged source of the female hormone – and the precise route by which it enters the childrens' food supply – remains in dispute.

What is not in doubt is the severity of the symptoms of the Puerto Rican children, some of whom came to the clinic of Dr Carmen A. Saenz de Rodriguez when they were less than two years old. Breast enlargement (in both boys and girls) was common. Little girls as young as four had grown pubic hair and their uteruses were developing. Dr Saenz, who has treated more than 600 young patients, reported that eighty-two per cent of the girls had ovarian cysts and there is fear that they will develop vaginal cancer when they grow older, as did the daughters of the American DES mothers.*

The U.S. Food and Drug Administration sent investigators in to establish the cause of the children's deformity, but they claim to have drawn a blank. Seventeen samples of meat and milk taken from local markets and tested back in the United States have proved strangely inconclusive. But Dr Saenz and her team, who deeply suspect this scant evidence, remain convinced that most of the children received huge doses of oestrogen by eating chicken which had been implanted with hexoestrol, one of the stilbenes compounds. Chicken is a popular food in the U.S.-controlled island, especially among the poor. And the

* *The Sunday Times*, 18 December 1983. For an excellent account of the events in Puerto Rico see Orville Schell's book, *Modern Meat* (Random House, 1984)

chicken producers had long been in the habit of caponizing birds – adding a hexoestrol tablet to the necks of males to give them a sudden burst of growth. Poor people, utilizing every scrap of food, commonly eat chicken necks in countries like Puerto Rico.

Although chicken is likely to have been the primary source of the Puerto Rican tragedy, other factors are still being considered. The island's natural vegetation, including some clovers, has an unusually high oestrogen content. This could have resulted in milk supplies being contaminated. It is also possible, although unlikely, that Puerto Ricans, for genetic reasons, are particularly susceptible to oestrogen exposure. Some researchers also believe that the large number of birth-control pill factories, sited in Puerto Rico for tax reasons by the pharmaceutical industry, could play a part in the puzzle. And Dr Ray Heitzman suggests an even more bizarre explanation: 'Puerto Rico is a poor, Catholic country where the church does not permit contraception. I have no proof but it is certainly possible that women have got a hold of stilbene pellets and are swallowing them. They make very good contraceptives, and the women may be taking them while they breast feed their infants to avoid getting pregnant again. High levels of the oestrogen could be absorbed by the nursing infants and the effects would only be seen later.'*

Stilbenes were finally banned in Britain in 1982 and the authorities contented themselves with the knowledge that the events in Italy had now passed and the perplexing problems in Puerto Rico occurred in a backward land thousands of miles away. Such misuse of drugs was not likely to occur any longer in the sensible democracies of the northern climates. That is why the events in Belgium in the spring of 1985 sent such perceptible shocks through the ranks of our drug regulators.

Belgium, Spring 1985

An illegal drugs ring, specializing in the sale of banned hormones including stilbenes, was uncovered in April after a six month investigation by Belgian police. The raid began with a tip off from a disgruntled drug trafficker. Three secret depots stocking the

* Dr Ray Heitzman, Institute for Research on Animal Diseases, interview with the author, 8 August 1985

hormones were uncovered in Belgium and fifteen people, including several veterinarians, livestock breeders and directors of a Luxembourg drugs company, were arrested pending the outcome of a full trial.

The Belgian police believed they had uncovered an illicit cache of growth-promoting drugs worth at least £50 million on the black market. Their enquiries led them into a network that covered Belgium, parts of northern France, Luxembourg and Holland.

Disclosure of the network caused deep embarrassment within the EEC, which, until then, had been arguing for only a partial ban on hormones. There was even some disquiet in Britain where all the legal growth hormones had been actively championed. The authorities on both sides of the Channel had always covered themselves with the caveat that hormones were safe so long as they were used properly. Here was clear proof that the hormone black market was still flourishing on a large scale and that illegal hormones were continuing to contaminate the 'wholesome' meat openly traded and sold within the EEC.

If such a sophisticated ring could be uncovered in one part of Europe, was it not likely that black market growth promoters, manufactured to get round the stilbenes ban, were being discreetly sold by other pirates throughout the EEC?

The Irish border

Some British farmers, when prepared to share a discreet joke, say they do not need a black market. They can get all the illicit drugs they want by making perfectly legal purchases south of the border in the Irish Republic, where control of chemical substances is the most lax among all the EEC countries.

In November 1984 journalists working for the British newspaper, *Farming News*, crossed into the Republic from Northern Ireland to see how easy it would be to buy hormones and antibiotics without prescription at five farm supply shops near the border. They had no trouble getting their scoop. At every shop they walked up to the counter without identification and were able to obtain at least some of the prescription medicines on their shopping list. They then carried them, without hindrance or inspection from customs officers more concerned about weapons, back into Northern Ireland.

Millions of pounds' worth of black market animal drugs are smuggled into Britain every year to be resold through disreputable farm merchants.

One of the drugs which the two journalists freely bought was chloramphenicol. Now we know what Dr Heitzman meant when he said it was 'the least prescribed, most used drug in modern (animal) medicine'. Cattle dealers are the main go-betweens in the traffic according to Mr Alan Davidson, deputy head of the UK Pharmaceutical Society's law division. 'Dealers on trips to Ireland or the Continent can easily buy huge volumes of prescription-only drugs, and smuggle them through UK customs. Many of the dealers conceal them in lorries which are literally impossible to check. The drugs are then sold at a huge profit to agricultural merchants.'*

It might be some comfort to think that this clandestine trade is confined to some extent by the English Channel and to illicit sorties across the Irish border.

The West Country, UK, 1985

But in July 1985 came certain proof that illegal drug trafficking was thriving undercover in Britain. Inspectors working for the Pharmaceutical Society swooped on an animal drugs ring operating in the western counties of England.†

Farmers from at least five counties had got to hear on the grapevine that prescription antibiotics could be bought directly from a network of agricultural merchants' shops that had spread covertly throughout the area. The farmers' greed was fuelled by at least two factors. Buying direct from the merchants enabled them to avoid both the professional fees of their vets and the prying questions that they might ask. As an extra bonus, the illegal drugs could be bought at prices far lower than those set officially by the drug companies. Explanations from within the pharmaceutical industry as to why the black market could undercut its own wholesale price to legitimate merchants were inventive, but not always convincing. An awful lot of legally made drugs seem to find themselves in illicit hands. According to Mr Davidson, his inspectors

* *Farming News*, 7 December 1984

† The *Guardian*, 2 August 1985

uncovered 'an Aladdin's cave of prescription antibiotics which the merchants had no legal right to possess or to sell to farmers'.

That the inspectors found the cache at all is a testament to their good luck and unflagging stamina. Although the illicit trade within Britain's borders is estimated to reach £3 million worth of animal drugs a year, the Pharmaceutical Society can only afford about twenty inspectors to police the whole country – and much of their time is rightfully spent inspecting chemist shops to prevent the illegal sale of narcotics and other dangerous drugs to addicts.* Mr Gordon Applebe, the Pharmaceutical Society's senior inspector, issued an immediate warning that 'the public could be at risk from meat and dairy products already contaminated by drug residues'. He knew that the West Country ring had been supplying farmers from a wide radius long before it was detected and investigations subsequently spread to the west Midlands. Admitting ruefully that his inspectors were hopelessly outnumbered, Mr Applebe said, 'We may just be scratching the surface of the problem.'

A string of successful prosecutions did succeed in October 1985 against merchants involved in the ring. Devizes magistrates' court heard that one of the convicted, Charles Richard Wyer who ran Animal Medicure Supplies, had sold £30,000 worth of drugs to fifty-four farmers and three other merchants during one six months period. He pleaded guilty to five specimen charges, asked for 254 similar offences to be considered and was fined a total of £7,500.† After the hearing Mr Applebe told reporters: 'This illegal supplying of prescription-only drugs is very widespread. This sort of business is done at the farm-gates all over the country.'

So where do the illegal supplies of drugs come from? The big pharmaceutical companies were not directly implicated in the West Country ring. There were suggestions that some of the antibiotics had originated in the Irish Republic where, as we have seen, it is as easy to buy a bag of drugs as it is to purchase a pint of stout. But several British veterinarians were also involved in the Pharmaceutical Society's investigations and evidence against them was passed to the Ministry of Agriculture. The clear implication is that veterinarians and their

* The number of inspectors was raised to twenty-four at the beginning of 1986

† *Daily Telegraph*, 29 October 1985 and the *Pharmaceutical Journal*, 2 November 1985

prescription pads are indispensable somewhere in the chain to act as go-betweens in the black market trade.

Such blatant corruption by 'bent' vets is only seldom revealed and hopefully, it is not very common. But this is not the seat of the problem. Many decent vets are subjected daily to commercial pressures to sign for prescriptions which allow unsafe doses of antibiotics to be fed to livestock on a wide scale. There is no secret about the pressure and the British Veterinary Association, which attempts to police the profession, has grown increasingly alarmed. Two of its leading members wrote an open letter to the *Veterinary Record* in February 1985 urging vets in the field to come forward with evidence. Neither of the authors had the least motive for being agitators. One, Dr David Wishart, actually works for Beecham, the large UK drugs firm which sells many veterinary drugs. But the facts were plain. Animal feed manufacturers make up regular consignments of calf milk laced with prescription doses of drugs like tetracycline. They also make up feed with drug levels that far exceed the approved dose. The *Veterinary Record* admitted: 'Sometimes the medicated feed has been manufactured and delivered so that (veterinary) practices are faced with a *fait accompli* when clients ask for a prescription for medication decided by others.'*

This open letter was final proof that the strict guidelines laid down by the Swann Committee to prevent resistant strains of bacteria from developing in animals and being passed on to man were being blatantly and systematically ignored.

Salmonella – the disease spread by drugs

South Dakota, U.S.A. – Wakefield, England

In the United States the Swann guidelines have never been adopted. They have been beaten by powerful lobbyists in Congress who speak for farmers' groups, drug companies and feed manufacturers. The U.S. Food and Drug Administration, which wants penicillin and tetracycline banned from animal feed, has been powerless against these allied commercial forces. The lobbyists have used the simple but effective ploy of getting Congress to write a special 'prohibition clause' into the FDA's annual budget, which prevents it from clamping down

* The *Veterinary Record*, 6 February 1985. See also 'Antibiotic Resistance in Salmonella', the *Veterinary Record*, 5 October 1985

on livestock drug use until there is conclusive proof of a link between animal antibiotics and human disease.

In January 1985 the Natural Resources Defense Council, the citizens' health pressure group, believed it had finally found 'conclusive proof' to beat the industry lobbyists and it submitted an 'imminent hazard' petition to the FDA during a full session of public hearings. The NRDC claimed that up to 300 deaths and 270,000 non-fatal cases of salmonella each year in the United States could be linked to the sub-therapeutic use of penicillin and tetracycline in animals.*

For evidence the NRDC cited new research which certainly came with blue ribbon credentials. It had been done by Dr Scott Holmberg at the Center for Disease Control in Atlanta (where Legionnaires' Disease was first identified and where research on a vaccine for AIDS has focused). His work was co-authored by Dr Thomas O'Brien from the Harvard Medical School and it was published in the *New England Journal of Medicine* (6 September 1984) which speaks with the same authority in America as the *Lancet* does in Britain. If further *gravitas* were needed, an editorial in the *Journal of Medicine* described the Holmberg-O'Brien study as 'compelling' and called upon the FDA to impose new controls on antibiotic growth promoters without delay.†

In its search for 'conclusive proof' the NRDC had surely struck gold. Human deaths have been caused, are being caused and will continue to be caused by resistant strains of bacteria which have been bred by the misuse of drugs in animal husbandry.

Holmberg and O'Brien had certainly done their homework. They looked at fifty separate salmonella outbreaks in the United States involving 2,000 victims. One quarter of all the bugs isolated after the outbreaks of food poisoning proved resistant to some antibiotics. A total of 312 people were infected with the resistant strains. Most of the victims lived, either because doctors finally found an antibiotic that would work, or simply because their own immune system ultimately overcame the infection after a long and painful struggle. But thirteen people died after the medicine-chest and their own defences were exhausted. Holmberg and O'Brien concluded that the death rate was twenty-one times higher than would have been expected in an outbreak of food poisoning caused by ordinary salmonella bacteria.

* *Animal Pharm*, 15 March 1985, p. 10

† *New Scientist*, 13 September 1984, pp. 3–4

Dr Holmberg then focused his microscope more narrowly to look at a specific outbreak of food poisoning in the Midwest during the winter of 1982–83 when eighteen people were infected and one died. All the victims had been infected with a strain of *Salmonella newport* which was resistant to ampicillin and tetracycline. Holmberg claimed he was able, by microscopic fingerprinting of the bacteria, to trace the salmonella strain back to a single South Dakota dairy farm where the herd had been fed sub-therapeutic doses of chlortetracycline antibiotics. By using computer records he claimed that thirteen of the eighteen patients had eaten hamburger that came from animals slaughtered in the dairy herd or from surrounding farms. And all the other patients, with one exception, were relatives of the thirteen primary victims. With even greater precision Holmberg also established that twelve of the thirteen dead of the 2,000 cases studied had been on a course of antibiotics given by their own doctors at the time of the outbreak. This was particularly damning evidence because it highlighted just how risky it could be for a consumer to swallow his doctor's prescription of antibiotics, and contaminated hamburger at the same time. The antibiotics had done a fine job of clearing the victim's system of many strains of harmless bacteria. Unfortunately, this had enabled the resistant bacteria eaten with the meat to run rampant with little competition through the victim's body.

As earnest researchers, Holmberg and O'Brien were only pursuing a line of scientific enquiry. But they ended up attacking a 'sacred cow' and they soon paid the price. Americans eat beef with one hand and salute the flag with the other. They do not really want to be told that devouring hamburgers can seriously damage their health, and farmers and the pharmaceutical industry who supply the meat would rather they did not believe it either.

Predictably, the lobbyists reacted quickly and fiercely to discredit the Holmberg-O'Brien study. The official sounding 'Animal Health Institute' described the research as 'flawed in many respects'. But this apparently impartial body turned out to be representing fifty-five drug companies that produce veterinary medicines.* Yet in November 1985 the U.S. Department of Health and Human Services rejected the NRDC petition and penicillin and tetracycline remain on sale without serious restriction to U.S. farmers.

* *Chemical Week*, 10 October 1984

The strongest argument against the Holmberg-O'Brien study was its failure to highlight the fact that the South Dakota herd had been given therapeutic doses of antibiotics for illness, in addition to the 'growth-promoting' doses combined with its feed. Which drugs were responsible for causing the *Salmonella newport* to become resistant? The distinction was crucial because supporters of growth promoters have always maintained that constant, low doses of antibiotics are not overpowering enough to force bacteria to mutate into new resistant strains to survive.

The American debate rested where the arguments in Britain had begun. How many herds and flocks of livestock are now receiving frequent high doses of antibiotics, signed for by veterinarians as preventative measures against disease, but which are actually used to boost growth illicitly and are causing strains of 'superbugs', dangerous to man, to develop?

The practice is so widespread in Britain that microbiologists at a salmonella conference in 1984 warned that a 'witch's cauldron' of new bugs were being bred which could cause 'an epidemic in food animals of a disease with the same potential for human illness as typhoid'.

Only weeks after these scientists made their prophesy, nineteen people lay dead at the Stanley Royde Psychiatric Hospital in Wakefield. All the victims were elderly patients who had been given a meal of cold, rare beef. Another 362 patients and members of staff were poisoned during the outbreak but they survived. Subsequent investigations were to establish that the beef had been left out in a warm room too long and salmonella infecting the meat had multiplied at a fearsome rate.

What remained unclear for many weeks was whether the salmonella bacteria in the rare beef were of a new, mutant and particularly virulent form – bred into a drug resistant super-strain. They were certainly members of the *Salmonella typhimurium* family, and as the *New Scientist* was to report: 'many outbreaks involving this type of bacteria display features of resistant strains'.

Meanwhile the authorities held their breath and to their intense relief their worst fears proved unfounded. The strain of *Salmonella typhimurium* which killed the patients was *not* drug resistant. The death toll was so high because the patients were already elderly and infirm and many could not be made to understand that they needed to swallow fluids to survive.

'If a resistant strain had got into a place like the Wakefield hospital, the disaster would have been many times greater,' said Professor Alan Linton of the Bristol University Medical School. He remains convinced that a drug resistant outbreak spread from animals is inevitable. 'It is likely to occur first in a hospital among very young children or the elderly who are already on antibiotics. It is just a matter of time really, and when it happens doctors will be able to do nothing, because they will not have any drugs left to combat the infection.'*

Professor Linton is no wide-eyed radical. In addition to his academic duties he has been chosen by the UK government to sit on its Veterinary Products Committee which studies all animal medicines before they are allowed to go on sale. He is also sensible enough to carry out real experiments involving human volunteers to prove his point that bacteria bred in livestock – including resistant strains – will insidiously get into man. When out shopping in Bristol he bought fifteen frozen chickens at random and gave them, three each, to five volunteers. Beforehand, Professor Linton tested the birds to see which resistant strains of bacteria they carried. He then tested the faecal samples of his volunteers who had cooked and eaten the chickens as they preferred, and discovered that at least one resistant strain of the common bacterium *E. coli* which had been found in the chickens, had colonized the gut of a volunteer.

The test was not particularly dangerous because *E. coli* is normally a pretty harmless creature. But what it proved was quite alarming. Microbes are by nature promiscuous. A resistant salmonella bacterium is quite capable of transferring its resistance, through tiny pieces of genetic material called R-plasmids, to bacteria of almost any other species. The process also works in reverse. A harmless *E. coli* bacterium carrying an R-plasmid can, once it is inside the human body, transfer its drug resistance to any colony of virulent bugs it happens to fancy.

Professor Linton offers another commonplace example. Doctors are finding it increasingly hard to treat urinary infections, particularly in women. Professor Linton suggests that resistant strains of *E. coli* which breed in the colon may in the toilet be wiped inadvertently in the direction of the genitals, causing subsequent infections which do not respond as they should to the common antibiotics. 'There is not a

* Interview with the author, 23 July 1985

shadow of a doubt that organisms, including resistant strains from animals do get through the food chain into humans,' he says.

So why do we not all die tomorrow or the next day from resistant strains of bacteria commonly eaten with our food? The reassuring answer is that most of us, in normal circumstances, are healthy enough to combat the infection with our own defences. But the stark answer is also that some of us – usually the weak and the infirm – are killed by 'superbugs' bred from animals – even if the precise cause of death is seldom spelt out in the *post mortem*.

According to Dr Bernard Rowe, 'There are bound to have been deaths in Britain from resistant strains of bacteria which have originated in livestock animals.'* At least forty people die each year in the UK from salmonella infection, but the deaths caused by the drug-

S.typhimurium phage 204c Britain

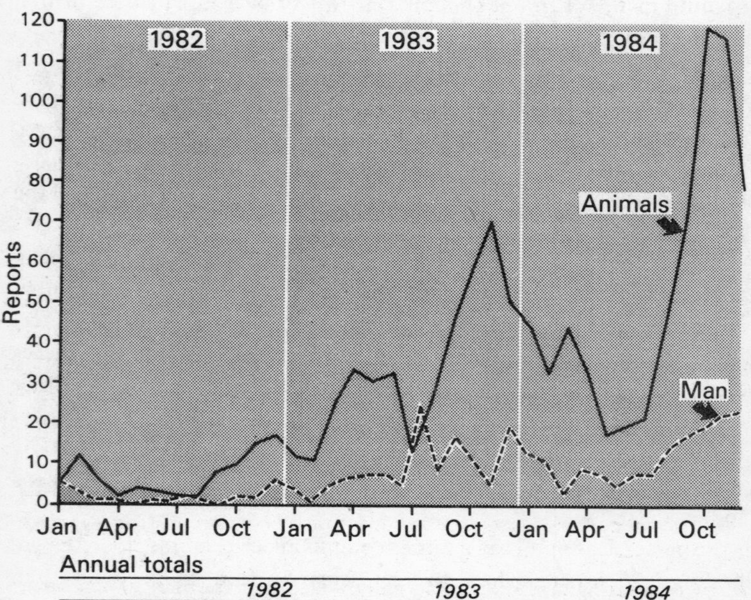

Annual totals

	1982	*1983*	*1984*
Animals	112	412	659
Man	26	81	127

Source: 'Animal Salmonellosis', *MAFF Annual Summaries*, 1984

* Interview with the author

resistant strains are not separately recorded. The authorities ought to be showing more concern since the number of reported food poisoning cases has more than doubled in Britain since 1970, and outbreaks of '204c' infection are rising dramatically.

Dr Rowe and his colleague, Dr John Threlfall, have systematically studied how *Salmonella typhimurium* type 204c, the drug resistant strain bred from excessive antibiotic use in cattle, has spread directly to man. The graph, which captures their alarming findings, shows just how closely infections in humans from the 'superbugs' track outbreaks of the disease in cattle.

Even if the young and healthy stand a better chance of surviving an attack from *Salmonella typhimurium* 204c, there is no immunity from the disease, and recovery, for the fortunate ones, can be slow and terrifying. A senior executive with a major drugs firm, whose anonymity must be protected, tells his own story about how he nearly lost his bride to a resistant strain of salmonella.

> Soon after I got married my wife got salmonella. We never knew where from. She ended up in Coppits Wood isolation hospital where *Salmonella typhimurium* was diagnosed. She was a young and healthy girl of twenty-six and she survived . . . but only after two very, very unpleasant weeks. When she tried to drink a glass of water it hit her stomach and just bounced straight out again. She was losing everything from diarrhoea and she was on constant intravenous drips. Thank God she was a strong and healthy woman – otherwise she would have become physically exhausted and she would have just died.*

That executive is to this day thankful to providence rather than to his own industry that his wife survived. No drug then devised could have saved her. And he privately admits that the pharmaceutical industry must share the blame for pushing more and more growth-promoting drugs into the hands of eager farmers.

Is it possible to get off the antibiotics treadmill? Sweden, in July 1985, became the first country to ban its farmers from feeding livestock a steady diet of antibiotic growth promoters. The Swedish Farmers' Meat Marketing Association actually welcomed the ban which came after strong consumer pressure. It said that attention to healthy husbandry was a more effective way of getting good growth than the use of antibiotics.

* Interview with the author, 3 September 1985

Sweden's bold decision came while the EEC was still debating whether to ban hormones from use as growth promoters. Certainly anxiety about the practice of pumping drugs into livestock is growing. But the use of pesticides on vegetable crops remains largely unchallenged, even though their dangers continue to grab the headlines.

California 1985

Watermelon is America's favourite summertime treat. The sweet red flesh tastes pure and should be protected by the melon's hard green rind. But in July 1985 more than 100 people in California and further north along the west coast were overcome with nausea, tremors, extreme sweating, blurred vision and diarrhoea after tucking into slices of the summer fruit.

California's health officials declared the epidemic an outbreak of pesticide poisoning and they ordered that ten million watermelons – one third of the entire U.S. crop – be destroyed. The chemical they found in the melons was Aldicarb, a pesticide so toxic that it ranks among the fifteen officially classified as 'poisonous' but still permitted for use in Britain.

Aldicarb is produced by Union Carbide, the U.S. chemicals combine, which moved swiftly to exonerate itself by claiming that farmers in California had been guilty of 'flagrant misapplication' of the insecticide.* Union Carbide could ill afford any more criticism of its conduct. The watermelon contamination occurred only months after the Bhopal disaster.

In California, Aldicarb is forbidden for use on fruit crops like watermelon. But some of the contaminated melons had come from fields where cotton had been grown in previous years. Aldicarb is permitted on cotton but Union Carbide insisted that the insecticide breaks down harmlessly in the soil within 100 days of application. The company therefore concluded that the wave of illness could only have been caused by the illicit use of Aldicarb by farmers directly on the watermelon crop. This theory did not explain why so many farmers would behave so stupidly in unison. Why did they all choose the summer of 1985 to dump an outlawed pesticide on a raw fruit crop so

* *Financial Times* and *Daily Telegraph*, 9 July 1985

redolent of the American way of life? Nevertheless, investigations by the California State Department of Food and Agriculture seemed subsequently to suggest that Union Carbide was probably right. 'We believe this was an intentional application of Aldicarb by some farmers,' said Mr Bob Rollins who was conducting investigations for the state. 'Offending farmers would face fines of $1,000 on each charge or six months in jail,' he said.*

A similar prosecution in Britain would be much harder to bring. For one thing, Aldicarb's legal limits are more tightly drawn in California. The insecticide is only permitted in fields where cotton, sugar beet and potatoes will be grown – and then only immediately after the seeds of the crop are sown. In Britain only the granulated form of Aldicarb, which may penetrate less deeply, is permitted. But it can be used on a much wider range of crops including salad vegetables like tomatoes which are seldom cooked and frequently unwashed when eaten. And it is not banned immediately after the seeds are planted but can be used up until six weeks before harvest time. Even if the farmer breaches British rules he is unlikely to be prosecuted, since the UK government has been reluctant to set maximum permissible pesticide residues on food. It is also unlikely to spend the kind of money necessary to enforce the regulations when, under pressure from the EEC, they are finally brought into force.

Nevertheless, it is only fair to say that proven cases of acute pesticide poisoning from crop residues, like the California watermelon affair, are rare throughout the developed world. Friends of the Earth, hoping to seize the headlines in Britain, went to the trouble and considerable expense of having tea samples analyzed in hopes of proving that our national beverage was contaminated with pesticide residues from countries like India and Sri Lanka where controls may be even worse. But the residue tests on tea proved nothing.

So far at least, scandals on the scale of the watermelon disaster require a genuine conspiracy involving a network of illicit chemicals or at least a concerted display of boneheaded stupidity and gross negligence by farmers who are normally loath to jeopardize their own livelihoods by threatening the lives of a large and visible group of consumers.

But pesticide residues, eaten slowly over a long period of time, may have a more insidious effect on our health which fails to attract the

* Interview with the author, 22 July 1985

attention of most doctors or of the media. Hormone and antibiotic residues may also be guilty of the same 'poisoning' by stealth. Some allergy experts, like Dr Jean Monro at the Nightingale Hospital in London and Dr Gina Shoentall at the Royal College of Veterinary Surgeons, believe that these chemicals do great harm even though the damage is fearsomely difficult to diagnose and measure. Their concerns are dismissed with polite indifference by the authorities in Whitehall who set food safety standards in Britain.

But the complacent stance of the authorities is vulnerable in two ways. First, it relies on the conventional scientific wisdom of our age. The experts who advise the government remain convinced that this assault of modern chemicals on our bodies is safe. But many of these experts also failed to measure the dangers of cigarette smoking, asbestos, whooping cough vaccines and stilbenes used during pregnancy (to name but a few) before the toll in human suffering and death became obvious. It is not easy to spot tragedies before they occur, and their reliance on the scientific assurances of industry has made them less than satisfactory watchdogs of the nation's health.

Second, the authorities only guarantee, with Olympian detachment, that the chemical residues we are forced to swallow are safe – *if* farmers, feed manufacturers and veterinarians use them strictly according to their instructions. Yet there is mounting evidence that pesticides, hormones and antibiotics are frequently and flagrantly misused.

4 / Down on the Farm: How Chemicals and Drugs Fuel the Food Machine

If the use and misuse of chemicals in the production of our food is a growing danger, then surely the farmer has a lot to answer for. But is he the prime culprit – or just the man left holding the bag of medicated animal feed or drum of pesticide?

In many cases today the farmer has little or no control over the food that reaches our table. Yes, farmers get a bad press: much of it they richly deserve. They are exceedingly prone to hide their true earnings under a bushel. And they bleat like lost sheep when the weather turns sour, insects attack, or cushy government subsidies are finally withdrawn after protracted complaint from irate, taxpaying consumers.

But not every farmer is laughing all the way to the bank in his Range Rover. Grain farmers are certainly the élite in Britain and EEC policy is geared to make them even richer. But livestock farmers can face a life of unstinting dedication to animal rearing with far less reward. Growing perishable fruit and vegetable crops can also be a hit-or-miss business with no guarantee of enough profit at the end of the season to keep the bank manager at bay.

These fears can make farmers cut corners with chemicals. Farmers may also hedge their bets by handing their output over to powerful commercial interests which will determine, in large measure, what chemicals must be used to guarantee yields and profits. Livestock and perishable crop farmers seldom enjoy a high intervention price from the Common Agricultural Policy for all that they produce. They cannot just dump their unwanted grain into EEC subsidized silos.

Instead, they can hand their destiny over to big corporations which will guarantee them a living . . . at a price. The price often demanded is that they follow strict regimes of production laid down by the supermarkets, food processors and feed manufacturers. Farmers can be told which hormones and antibiotics to feed to their livestock and what pesticides they must apply to their crops. The food machine is geared to ensure that the crop is plucked and the animal is culled so that they end up on the supermarket shelf in the quickest, most profitable time.

Agriculture is the biggest single industry in Britain, the EEC and the United States. Putting food into our affluent mouths is too big a business these days to leave to chance. Fifteen years ago this kind of 'contract farming' was still in its infancy – but even then, a UK government report remarked that farmers were handing over *de facto* control of their output and lowering their status to 'little more than that of an employee'.* Today most of the poultry, eggs, pork, bacon and crops like beans, peas and carrots are grown to the order and specification of the food industry.

'The farmer is over a barrel when a large supermarket says to him: either you get us such and such at a certain time and in this condition, or we will go elsewhere,' said Mr Ian Dalzell, a marketing executive with the National Farmers' Union. 'This may force the farmer to use chemicals or hormones or other particular treatment. In some cases they may actually be specified.'

Often the farmer is squeezed between two millstones – the supermarket retailer at one end and the supplier of feed and drugs at the other. Sainsbury, for example, does not leave putting pork and bacon on its shelves to chance. The supermarket has a major commercial tie-up with Pauls, one of Britain's biggest animal feed producers, through a jointly owned company with the pastoral name of Breckland Farms Ltd. Breckland decides which blends of Pauls' animal feeds and drug additives will be fed to preselected piglets. These will be handed to the farmer on a contract basis.

After the pigs are reared according to the strict production schedule they will be taken, in many cases, to be processed for slaughtering and dressing at another company in which Sainsbury has a direct interest, Haverhill Meat Products. From the piglet's first oink to its last gasp Sainsbury has exercised control over its life – including which chemicals have been added to its diet.

Trying to unravel the complex web of contract farming in Britain is

* *Contract Farming: Report of the Committee of Inquiry on Contract Farming* (HMSO Cmnd. 5099, October 1972), p. 9

not a simple task. Questions put to the Ministry of Agriculture about who controls what in British farming draw a blank. Either the Ministry does not know or (given its love of official secrecy) it is not telling. The big supermarkets, feed manufacturers and food processors can be equally unhelpful, claiming that they do not want to divulge any crucial information to their competitors.

What is clear is that in all areas of farming control is becoming more and more concentrated into fewer and fewer hands. Nowhere is this more evident than in poultry and egg production. Nine out of ten chickens we eat in the UK are raised in flocks of at least 20,000 birds. Small poultry farmers who own less than 1,000 birds have virtually no impact on the market. The same is true for eggs. Although 45,000 people in Britain still keep laying flocks of less than 500 birds, they produce just three per cent of the eggs we eat. Nine out of ten eggs come from laying flocks of at least 10,000 birds.*

The National Farmers' Union, which has a vested interest in protecting the dwindling independence of its shrinking membership, has made the greatest effort to study the impact of contract farming. The following table, drawn up with the NFU's help, gives a rough guide to some of the direct control now exercised by the wider food industry on British farming:

Product	*Amount Controlled by Contract Farming in UK*
Poultry	60%
Eggs	35–50%
Pigs (pork & bacon)	40–50%
Peas	95%
Green beans	95%
Carrots	25–30%

Estimates for beef, veal and lamb production are even harder to come by. Farmers, it seems, do not want even their own union to know who they are doing business with. Milk and cheese production is, of course, almost exclusively contracted to the Milk Marketing Board.

Defenders of contract farming argue that it benefits consumers who

* 'Annual Guide to the Poultry Industries of the EEC,' *Poultry World*, 19 Sept 1985

are constantly demanding cheaper, more attractive looking food which appears to have rolled off an assembly line. A constant supply of uniform, cheap food cannot be produced without the magic wand of chemical agriculture. Pig and poultry farmers, or those who grow the peas and beans for the big food processors like Bird's Eye (Unilever) have little choice but to sign up as contract employees. If they buck the system they face the risk of having their small amount of produce undercut by the output from the big feed manufacturers, food processors and retailers who dominate the market place.

The one company which has most rapidly refined contract farming into a production line process is Hillsdown Holdings. In a single decade it has sprung from nowhere to a commanding position in the British food industry. Hillsdown's 'Daylay' division is the biggest egg producer in the UK supplying twenty-two million eggs to the market every week. Through its 'Buxted' brand Hillsdown is also Britain's largest chicken producer (one million a week) and it is also number two in turkeys.

Despite this High Street prominence, Hillsdown prefers to keep a very low profile. The company's joint-chairmen, Mr David Thompson and Mr Harry Solomon, refuse personal interviews and, unusually, avoid having their photographs taken by the national newspapers.

Yet Hillsdown's influence does not end with just poultry and eggs. Hillsdown has also acquired FMC, the biggest meat wholesaler in Europe with forty-eight abattoirs handling cattle, sheep and pigs dotted around the country. For good measure the company has also become Britain's biggest bacon-curer through its Harris division and it also owns Lockwoods and Smedleys, the country's largest canner of fruits and vegetables.

Hillsdown is not just big, it is highly profitable. Its profits soared from £1.4 million in 1979 to £18.8 million in 1984 and investors in the City of London rushed to buy a piece of the company when its shares were floated on the stock market in 1985. Said one admiring analyst: 'If there is nine pence to be made, Hillsdown will find a way to squeeze out a shilling.'

One of the things that attracted investors is Hillsdown's ability to guarantee profits by controlling food production from start to finish. Feed and chemical additives make up seventy per cent of the cost of raising intensive livestock. Hillsdown solves this problem by owning Nitrovit (one of the big five animal feed producers in the UK) and the related company, Nutrikem, which provides the drug additives. Nitrovit produces 40,000 pigs on its own farms every year. But to ensure an outlet for its medicated feed it also puts 160,000 piglets out to be raised on contract by other farmers. 'We specify the breed, the feed and the growth-promoting antibiotics that the contracted farmers must use,' said David Speight, the director of nutrition and research at Hillsdown's Nitrovit division. Any pig farmer unhappy about feeding his contract animals this regime of drugs could discuss his concern with the company. 'But I suspect he would have a problem meeting our growth targets without them.'

There might also be other complications. Regular orders for pigmeat from Nitrovit and its contracted farmers are placed by supermarkets like Sainsbury and by Marks and Spencer. 'They are very, very specific. They will tell us exactly how much growth promoter they want to be used – they dictate it,' said David Speight. 'The supermarket is increasingly calling the tune in quite a big way.'

Hillsdown believes that medicated meat manufacture is the only way to ensure healthy animals, quick growth and low cost. All of its own pigs receive one of the approved growth-promoting antibiotics like Avotan (avoparcin) or Eskalin (virginiamycin). And despite high standards of husbandry which have gained respect throughout the industry, Hillsdown still resorts to the more potent prescription antibiotics on a regular basis when disease, or the threat of disease, is expected. 'We give our pigs prophylactic doses of these antibiotics less than ten per cent of the time. We do not have any option but to use antibiotics that are also used in human medicine,' said David Speight. The Buxted poultry division operates in a similar fashion. Birds destined for the dinner table receive a steady dose of the approved growth-promoting antibiotic, zinc bacitracin, in their diet. But higher doses of the human antibiotics are used, says the company, 'only in the face of a known disease challenge'.

Lockwoods and Smedleys, the Hillsdown canners, buy in all their fruits and vegetables from contracted growers. 'We certainly lay down guidelines on the sprays, pesticides and fungicides which we are prepared to have used on the crops we buy and the times that they can be applied,' said Dr Malcolm Crouchman, a senior executive with the company. But is there any way to know whether the contracted farmers, who face tight harvesting schedules, always stick to the rules? Dr Crouchman admits: 'Nobody is checking our fruits and vegetables for pesticide residues on a controlled basis.' Why not? 'As a consumer, of course, I would say there should be systematic residue checks. But as a manufacturer I can see the problems it would cause us.' At the height of the season Lockwoods buys in 100 tonnes of peas a day and the canneries are running full tilt. No one is going to keep the machines idle while batches are sent off to the analyst for routine sampling.

Hillsdown's conduct has been highlighted because its activities spread widest and deepest throughout the UK food industry. It appears to pay more attention to the hazards of chemicals and drugs in agriculture than other farmers and producers, and there is evidence that safeguards at Hillsdown are among the best in the industry. For example, Dr Crouchman says that dubious fruit and vegetable samples are sent to the government-run Camden food research station if a problem is suspected. And back at Buxted Poultry they claim to use no regular antibiotics of any kind in egg production. Furthermore, all drugs are withdrawn from table birds seven days before slaughter – a costly practice which exceeds government regulations.

The inescapable fact is that the consumer, led by his supermarket, has come to expect the kind of low cost, apparently attractive food which can only be produced by resorting to the chemistry set of antibiotics, hormones and pesticides.

But does such low cost, industrial agriculture always produce safe food, free from chemical residues which may prove harmful? Apparently not. Dr John Walton is a lecturer in Veterinary Preventative Medicine at the University of Liverpool. He could hardly be described as an extreme critic of the use of drugs in food production. In fact, the Association of the British Pharmaceutical Industry was so approving of his views that it reprinted one of his recent research articles. But Dr Walton fears that the demand for cheap food inevitably means the development of fast-growing but disease-prone animals which cannot survive the cramped assembly line without recourse to more and more drugs.

Over the years new animal hybrids have been developed, highly productive strains of animals and birds have been developed and foreign breeds have been introduced into the UK often with the sole intention of increasing productivity and carcass quality or reducing 'days to market'. Unfortunately, very little attention appears to have been given to selecting strains of animals for resistance to disease or durability in modern housing environments. This has resulted in some cases in an increased prevalence of disease which has to be constantly treated or prevented.*

So what does the contract farmer do when his herd or flock shows signs of possible disease which could wipe out not only this year's earnings, but also the prospect of having a contract next year? The answer, according to one Ministry of Agriculture veterinarian (whose identity must be protected) is that the farmer is sorely tempted to reach for the bag of drugs. 'If one or two animals in a herd get sick it's reasonable to prescribe a therapeutic dose for them. But vets are under constant pressure from the drug companies, from adverts in the farming press and from farmers themselves. The easiest thing is to prescribe a preventative dose for the whole herd – and then keep them on the antibiotics indefinitely.'

The problem is best illustrated by the dilemma of farmer John Walton (no relation to the Liverpool veterinary lecturer of the same name). He is a hospitable and unpretentious man in his middle fifties who rears pigs on thirty-five cramped acres near Wrotham in Kent. Pig farming has failed to make John Walton rich. Cereal farmers nearby may trade in their Range Rovers every year, but he and his wife Annie make do with a five-year-old Morris Marina estate. Yet by UK standards he is a medium-to-large pig farmer. His collection of somewhat down-at-heel sheds and breeding units house 175 sows, ten eagerly salivating boars and a total of 3,500 squealing piglets which are grown for slaughter every year. Most of John Walton's pigs go to a large supermarket via a local co-op, South East Pig Producers, which he was instrumental in establishing in 1981. His daughter, Gillian, manages the co-op for several other local pig farmers as well.

As I step into my borrowed green wellies for the tour of the farm I am immediately struck by John Walton's candour and recognition that his economic fate has ceased to be within his own control. 'I wish I

* Walton, John R., *Antibiotics, Animals, Meat and Milk* (1982, reprinted by The Association of the British Pharmaceutical Industry), p. 82

could stand here and tell you that I rank among the top ten per cent of pig farmers in this country in terms of efficiency . . . but I don't. I am definitely in the top third,' he says with hard won pride. 'But I have been held back by a disease problem which has been very hard to get rid of.'

The disease which has plagued the Walton farm is scours, the all too common but highly virulent form of dysentery (caused by salmonella and other organisms) to which both pigs and young cattle are particularly prone. Its symptoms include virulent infection and bloody diarrhoea. The disease can spread rapidly and threaten the entire herd if it is left untreated.

During the height of the scours outbreak John Walton was advised to use Fortigro-S, the prescription antibiotic which the drug company, Pfizer, was forced subsequently to withdraw.

By the time of my visit there had been no sign of scours for six months and John Walton was crossing his fingers that he had finally got the disease licked. But a quarter of his herd – typically the more susceptible younger piglets – were still receiving preventative doses of other prescription antibiotics that remain legally on sale.

Rural artists in search of pleasant pastoral landscapes seldom choose pig farms. They are not appealing places. The pigs I saw on the Walton farm looked healthy enough under their circumstances and all drugs, I was assured, are withdrawn for the approved time prior to the animals being carted away for slaughter.

But after more than thirty years in the business John Walton has taken a resigned, if increasingly sceptical view about the constant need to use drugs in modern pig rearing. His own father was a research chemist for the Wellcome Foundation drug company and his twenty-four-year-old son Robert now works for BOCM-Silcox, the division of Unilever which supplies medicated feed to farms all over Britain.

Did John Walton think he has more disease problems now than he did when he started out with just a few sows on the same farm in 1949? 'Yes, I suppose I do,' he said. But did he think that steady resort to antibiotics, both as growth promoters and for disease prevention, had aggravated the problems by causing resistant strains of bugs to develop? 'No, I don't think so. The real problem is that I have had to increase the size of my herd constantly just to make a living. I have to push my pigs to the limit.'

The economics of pig rearing are pretty stark. Twenty-five years ago John Walton received around £20 for every pig he sold. Even though he had fewer sows which bred fewer piglets over a slower cycle, the income he received made him feel affluent and confident of his future. Certainly in comparison to the average national wage (in those days less than £1,000), he was a rich man. Today John Walton receives only £60 for every pig he sells. That three-fold increase has left him relatively a lot poorer because the average national wage has risen ten-fold to £10,000 over the same period. He has been knocked down the economic and social ladder by inflation and other people's rising incomes. Overcapacity in the industry, the absence of lavish price subsidies from the EEC and the power of retailers to drive down pork and bacon prices have meant that John Walton earns less and less for every pig he raises. To survive he has to squeeze an ever bigger herd into the same thirty-five acres and to fight diseases when they spread with the available drugs at his command. He admits that better buildings to house his pigs might help reduce disease naturally. But they cost a lot of money – the kind of money which John Walton can no longer ask his bank manager to lend.

Intensive livestock farmers are not the only people under pressure in the countryside. As their numbers shrink and production units get bigger in a search for economies of scale, the veterinarian also finds himself with fewer potential customers. Vets who have to tout for business have to meet the demands of their dwindling clientele if they want to stay in practice. The ethical dilemma can be great. Veterinarians may feel decidedly uneasy about dosing so many animals with drugs. But they are self-employed businessmen who cannot make a living without the money they earn from prescribing and directly dispensing drugs. Besides, refusing to sign a prescription demanded by a farmer whose medicated feed has already arrived, may be as pointless as it is high-minded. 'Farmers who are working to contract do not always have to consult their local vet when they need prescription drugs,' said the Ministry of Agriculture vet. 'Why should they when they can call in a vet who is on the payroll as a consultant to the feed company or food processor who has contracted them? His professional services are free.'

In theory at least, chemical agriculture is supposed to be safe for the consumer. But defenders of the system in government and industry are always careful to abdicate responsibility. They insist on adding the health warning that these potentially hazardous compounds must always be applied in the correct amounts and at the right time 'by persons fully conversant with their pharmacological properties and thus able to avoid such problems as tissue residues and certain direct toxic effects'.* These are fine words. But is there any real hope that lofty guidelines like these, drawn up in isolation by learned scientific committees, and laid down by bureaucrats in Whitehall will be carried out with strict obedience down on the farm? Surely not. Few farmers have degrees in pharmacology. And sadly, many have every incentive to break the rules, and little fear that their petty misdemeanours, which collectively create a hazard, will ever be apprehended.

By the way of a simple example let us trace how a calf, taken from its dairy-cow mother when only a day or two old, is reared for market before it ends up as beef on our dinner table.

It is well known that calves lead a miserable, short life. They are weaned and separated from their mothers before any natural bond, which will upset the valuable milk-producing mother when severed, can form. Her milk is rich in antibodies that would have given the tiny calf some protection from infection as it begins its orphan journey. But the farmer earns too much from the liquid milk, sold at the high EEC intervention price, to waste it by allowing it to go down the calf's throat. Instead the calf is transported, distressed and frightened, to the cattle-rearer who feeds it on medicated milk powder bought at a subsidized price from the stored dry-milk mountain that the EEC is eager to reduce – which is made, of course, from the same liquid milk the calf was not allowed to drink.

The antibiotics added to the milk powder will ward off some infections and make the calf grow at an unnatural, but profitable speed. Nothing illegal has taken place. But the young animal is forced into a confined space occupied by many other calves which are likely to have brought some infections with them. Taking no risks, the cattle-rearer may want to add prescription doses of antibiotics to the animals' feed to ensure that any potential infection does not spread. Often a compliant vet is at hand to sign for the drugs which the farmer, having read glossy adverts in the farming press, is only too eager to buy. If a vulnerable veterinarian is not available, or if the farmer is particularly

* Walton, John R., op. cit.

greedy or vulnerable he can probably find his way to the black market where, as we have already seen, drugs from Ireland or even of domestic manufacture, can be acquired at a lucrative discount even when they have been banned from legal sale. The twenty-odd inspectors of the Pharmaceutical Society who must police the entire country are unlikely to intrude.

The cattle-rearer may dispose of some or all of his herd and the growing calf may pass through several markets and be exposed each time to resistant salmonella bacteria or other infections. When the animal finds its final resting place with the cattle-finisher it may be castrated and implanted with a growth-promoting hormone before it is put out to grass for fattening. The hormone pellets in its ear will only add to the growth-promoting dose of antibiotics it will continue to receive in supplementary feed.

If the farmer is careful he will withdraw any therapeutic antibiotics and synthetic hormones for the required period before he allows the animal to be slaughtered. If his husbandry is less exact there is little to prevent him from ignoring the law. The chance of anyone from the ministry or the supermarket testing the animal's meat for drug residues will be remote in the extreme.

'A farmer has only one aim and that is to get more meat on the hook,' said Joy Wingfield, who is one of the twenty Pharmaceutical Society inspectors. 'If an animal looks to be in a poor state lots of farmers will just push more antibiotics into it in hopes of getting it up to weight. But there isn't a cat's chance in hell of such a farmer getting caught if he overuses antibiotics and hormones or ignores the withdrawal periods and leaves high residues in the meat he produces.'

The trouble is that farmers have every incentive to overuse drugs and no compelling reason to stop. Commercial pressures, whether they are dictated by a contracting paymaster or by the sheer rigours of the market-place, insist that they boost the weight of their livestock as quickly as they can.

Growers of fruits and vegetables have no reason to behave differently. When supermarkets like Tesco, Safeway, Sainsbury and Marks and Spencer contract directly with farmers for supplies of fruits and vegetables they theoretically require that the fresh produce be delivered free from chemical residues or 'taint'. They can hardly check every consignment. The major supermarkets do make some checks by sending occasional samples to laboratories – but they are often loath even to admit to this kind of scrutiny. Presumably, an open admission that they do some haphazard testing might alarm customers who like to believe that all food they purchase in their polished-floored supermarkets is pristine.

But Peter Atkins, the Kent farmer who regularly supplies the major supermarkets with fruit and vegetables, accepts that he cannot avoid leaving some pesticide residues on his crops if he is also to meet the stringent 'appearance' demands set by the big retailers.

'The supermarkets tell me exactly how long, how wide and what colour my crops are supposed to be. To meet these specifications we have to use chemicals,' he said. 'As I have told you, the biggest complaint I get is about caterpillar shit on my cauliflowers and greens. Marks and Spencer sends me copies of customers' complaints. They take some reading to be believed.'

To avoid any hint that insects ever walked across (or did something worse to) his broccoli, cauliflowers, sprouts and cabbage, Mr Atkins is forced to use insecticides right up to the limit of the minimum withdrawal period specified on the label. 'I would much prefer to reduce the chemicals I use because the odd few holes in something render it no different when it is in the pot, but I cannot because Marks and Spencer and Sainsbury will reject it for insect damage. I'd like to cross my heart and say that I never feed anybody pesticide residues, but I do not believe that this would be true.' So is the fresh, clean good-value-for-money image which the supermarkets try so hard to promote always justified?

'Consumers who buy in the big supermarket could face a slightly bigger risk because the big multiples put such pressure on us for cosmetically perfect crops which taste no bloody different, but look pretty in the pack, and so the risk from residues – because the farmer must spray right up to the approved withdrawal time – is there,' Mr Atkins added.

Officially, the authorities insist that the pesticide residues commonly found in food in supermarkets and other shops are harmless. And the pesticide companies insist that they test and retest their products to ensure that the margin for safety is at least 100 times greater than the exposure that anyone could conceivably receive from their chemicals.

But is the consumer still being misled about what residues are in his food and why they are put there? ICI, Britain's biggest pesticide manufacturer and the fifth biggest in the world, prides itself on its safety record. Mr Chris Major, formerly the director of the British Agrochemicals Association and now an ICI spokesman, believes that the food industry falsely parades an advertising image which beguiles customers into thinking that modern yield-boosting chemicals never get near their food.

'You only have to look at all those quaint and antiquated cart-horses going across our television screens advertising anything from beer to cheese,' he said. But this ancient agrarian imagery redolent of Constable's 'Haywain' is deeply deceiving. 'It gives the impression that these high quality foods being advertised can somehow be produced at the price we enjoy without modern technology – and some of that technology is, of course pesticides and other chemicals.' Since chemicals are here to stay Mr Major believes it is time that the food industry stops conning the public and begins through its advertising to admit the necessity and extol the virtues of food and life with pesticides.

Consumers certainly deserve complete candour about the modern methods used to get food on their table. If farmers never used faulty spray nozzles, never misread complex labels, or never put too much pesticides on their crops too late by design, to boost their yields, then perhaps most of these toxic chemicals, applied strictly according to instruction, might be safe for most consumers. But the reality is very different.

As we have already seen, fully one third of all our fruits and vegetables, as tested by the Association of Public Analysts, contained detectable levels of pesticide residues. The authorities and the pesticide companies insist that this is a false alarm. Detectable levels are so minute, even if they can be detected by the sophisticated equipment of today, that they cannot possibly do any harm to anyone.

Perhaps. But British detection equipment is hardly the most accurate in the world. Government expenditure cuts have left the public analysts without the latest devices which could make detection of minute residues of toxic substances more certain. In the United States there are some labs that can find residues which totally escape British analysts' detection.

Even so, the public analysts found that ten per cent of all fruits sampled and twenty per cent of all vegetables contained pesticide residues above the 'reporting' level set by the Ministry of Agriculture itself. Is this just a false alarm too? And what about the separate test by the analysts which revealed that one in ten samples of apples, brussel sprouts, cabbages, lettuces and mushrooms contained the same excess levels of DDT?

One ministry official who cannot be named was concerned and candid enough about current safety standards to remark: 'Ministers tend to overstate the level of security we are at. What the officials will say on the record is that the residues are *mostly* below international standards and declining. They are talking about some of the organo-chlorines (like DDT) and these jolly well should be declining since many of them have been banned. But this does not mean that the residue situation in Britain gives no cause for concern.'

5 / In the Boardroom: The Abuse of Power by the Chemical and Drug Companies

The drugs and pesticides used in agriculture are made by chemical and pharmaceutical companies which rank among the most profitable and influential corporations in the world.

The products made by these chemical combines infiltrate every part of the globe and every other industry. In good years the bigger corporations like I C I earn profits in excess of £1 billion. They make their money by utilizing life's simple building blocks – salt, water, hydrocarbons and lower organisms – to produce every conceivable saleable substance from toothpaste and chewing gum to space-age plastics and life-saving drugs.

Corporations this big, which touch all our lives so intimately and can even wield the power of life and death, are bound to attract political and social controversy. Consumers reasonably expect to know who is responsible for the weird and wonderful compounds with unpronounceable names which invade the air they breathe, the articles they touch and the food they eat. Just as surely, these corporations are determined to use their power and wealth to promote and defend their own interests when they are attacked as greedy, arrogant and irresponsible by consumer groups, environmentalists and politicians. Dr Louis v. Planta of Ciba-Geigy makes an easy target of himself when he apparently suggests that genuine public concern is being manipulated as a vendetta against capitalism by subversive organizations like consumer groups, Third World campaigners and the World Council of Churches. Few leaders of the chemicals and pharmaceutical industries can be quite so extreme in their outlook. But many of them – men and women genuinely proud of their companies' achievements – do feel unfairly vilified. They believe the media and the public take their successful products for granted and only choose to be outraged when a drug or chemical slips through the safety net and proves a killer rather than a cure.

On several occasions during research for this book chemical company executives have asked with exasperation why no one has set down in detail how many lives have been saved, and how the real cost of life's little luxuries like roast pork and chicken have been significantly reduced by the genius of their laboratories and factories. It is quite true that drugs save many lives and that intensive farming, which grows more crops per acre and packs more pigs and poultry to the pen, could not exist without modern pesticides and veterinary medicines.

But public altruism is not the prime motivator of chemical companies. Profit is. Developing new drugs and pesticides is extremely expensive. Increasing concern about safety means that a decade can elapse before a patented new compound meets all the tests of the authorities and is finally allowed on the market. This research, scrutiny and delay can cost up to £50 million for a new drug and £20 million for a new pesticide. Frequently the money spent is tossed down the corporate drain, either because the product does not work or because, at the eleventh hour, it fails a safety hurdle. If the stakes of this gamble are high, so can be the rewards. Sales of pesticides (excluding Communist countries) now exceed $15 billion a year and nearly eighty per cent of the total is bought and used in the developed world. A single drug, like the anti-ulcer agent, Tagamet, earns profits for SmithKline, the U.S. pharmaceutical company, of an estimated $300 million a year. Profits from veterinary drugs are smaller. But many of these drugs, like antibiotics, were originally developed for the human market, and the companies are eager to recoup these research costs with profits made from farmers. Sales of veterinary drugs reached $2.5 billion in the United States and $245 million in Great Britain in 1983. By 1990 worldwide sales of veterinary medicines, of which growth promoters are the biggest category, should exceed $10.5 billion.

With this kind of money at stake, it is not surprising that the biggest and most powerful chemical conglomerates tend to dominate the manufacture of pesticides and drugs, both human and veterinary. Only they can afford to roll the dice, to scoop the big pool, and now, more than ever, they need to. For a generation the chemicals industry grew fat on the money it could make from its investment in the manufacture of bulk chemicals, fertilizers and plastics. Now profits from these heavy industries, squeezed by the oil crisis and world recession, have been scarce. All the big corporations, like DuPont, ICI and Hoechst are pouring their new investment into agriculture and health care where the market, and the proceeds, are still expanding. When ICI earned profits of over £1 billion in 1984, a third of the group's total earnings came from drugs and pesticides.

With so much at stake, the giants of the chemical industry are prepared to defend their interests and, when they think necessary, to throw their weight around.

The Swiss firm, Hoffmann-La Roche, which is best known for its tranquillizers and vitamin tablets, is also the fourth biggest veterinary drug manufacturer in the world. When, in 1973, a former executive of the company, Mr Stanley Adams, decided to report Roche's illegal price-fixing activities to the European Commission, he was thrown into a Swiss jail and charged with industrial espionage, a crime tantamount to treason against the Swiss state. His young Swiss wife committed suicide after she was told by her police interrogators that her husband would be put behind bars for twenty years. Hoffmann-La Roche was ultimately found guilty of price fixing by the European Commission and fined £240,000. Mr Adams subsequently escaped the full penalty of Swiss justice but his life was ruined. He finally won substantial compensation against the European Commission in November 1985.

As we have already seen (Chapter two), Ciba-Geigy, another of the big Swiss firms, chose to test its unproven pesticide, Galecron, by spraying it from an airplane on Egyptian children.

The ethical standards of the big drug and chemical companies tend to decline in direct proportion to the distance their activities are from the glare of the western media. The U.S. drugs firm G. D. Searle (which is now part of Monsanto) was barred by the American health authorities from allowing its anti-diarrhoea drug, Lomotil, to be prescribed by doctors to children under the age of two in the United States. The drug, which acts as a potent chemical 'plug', was judged to be unsafe when used on infants. But this ruling did not prevent Searle from continuing to sell profitable batches of Lomotil in Africa and other developing countries, where it was offered without prescription, warnings or age restrictions, to unsuspecting parents hoping to cure their children of dysentery. Mr Charles Medawar, the health activist who campaigned against Lomotil's unrestricted sale, publicized how the misuse of the drug contributed to the death of many children in the Third World.* They would have stood a much better chance of surviving if their bowels had been left 'unplugged' and they had been given plenty of fluids until the infection ran its natural course.

ICI, the UK chemicals and drugs combine, prides itself on its social responsibility. Yet even ICI has been guilty of promoting Anapolon, the anabolic steroid, as a wonder cure for malnourished children in Bangladesh. Leaflets advertising the sale of the growth hormone to poor parents in the country were withdrawn only after Oxfam publicly highlighted the abuse – and even then, the drug remained discreetly on sale.†

So much has been written about the Bhopal tragedy in India that it seems almost unnecessary to record that Union Carbide failed, for more than a year after the deadly leak of pesticide intermediates from its plant, to pay any substantial compensation to the thousands of victims of the disaster during the months when they needed help the most.

If the big corporations tend to show the least social responsibility abroad, they can still behave with arrogant stupidity even in their own home towns.

In July 1985 the Dow Chemical company grew exasperated when protestors, working for the environmental group Greenpeace, repeated an attempt to plug up the main effluent pipe at its headquarters plant in Midland, Michigan, where pesticides contaminated with dioxin are produced. Local policemen, suitably clad in frogmen's wet suits, arrested three of the protestors midstream and took them to jail, where they were detained for three days on a trespass charge while Greenpeace was forced to find $30,000 for bail. While in jail the protestors, including twenty-eight-year-old Melissa Ortquist, a woman of previously good name, were forced to submit to blood tests. Dow illictly obtained the results of the blood tests which erroneously indicated that Miss Ortquist, the daughter of a Protestant minister, had syphilis. Subsequent independent tests proved she had never had the disease. But in the meantime an executive of Dow, acting on the company's instructions, had already telephoned a local Greenpeace supporter to tell her the news in a manner apparently designed to discredit the young woman and Greenpeace's activities. After a storm of public protest and

* Medawar, Charles, and Freese, Barbara, *Drug Diplomacy* (Social Audit, 1982)
† The *Guardian*, 19 January 1983

international publicity Dow was finally forced to apologize for its 'serious error of judgement', and the court subsequently acquitted the three Greenpeace defendants on the trespass charge.

Committed critics of the chemicals industry believe that all the major corporations are frequently guilty of similar behaviour, but are merely better at keeping such tactics covert. This is probably unfair. Certainly many executives within the industry despair privately that such unpleasant episodes bring their own corporations, indiscriminately, into disrepute.

Sometimes, however, the large firms act publicly and in unison in ways that draw criticism. A group of American drug companies, angered by the treatment their UK subsidiaries were receiving from the Conservative government, banded together in 1985 in an attempt to organize a £500,000 lobbying campaign aimed at influencing ministers, important back-bench MPs and Whitehall officials who regulate drug prices. Few of the companies seemed to appreciate that public disclosure of this fund might lay them open to the accusation that they were cynically prepared to buy their way into the corridors of power, in a crude attempt to protect their profits. Then again, why should the Americans understand such subtleties when British drug companies, operating under the umbrella of the Association of the British Pharmaceutical Industry, had behaved just as ineptly only months before? Right in the middle of its battle with the government over drug prices and profits the ABPI headhunted a new director, Dr John Griffin. And who was Dr Griffin? None other than the senior civil servant in the Medicines Division of the Department of Health whose job had been to vet new drugs for their safety and efficacy before they were allowed on the market. In many eyes the government's foremost gamekeeper had suddenly become the drug industry's prized poacher.

Eyebrows may have been raised, but in Britain the drug industry has had a long tradition of making sure that it has kept on good terms with politicians and other policy makers. Both Glaxo and Beecham, two big British drug companies heavily involved in veterinary medicines, have donated generously to the Conservative Party in recent years. Over a similar period at least nine Members of Parliament have served as directors or consultants to major drug companies.* And at least four

* Braithwaite, J., *Corporate Crime in the Pharmaceutical Industry* (Routledge and Kegan Paul, 1984), p. 300

serving MPs have admitted to having been drug company consultants in the Parliamentary Register of Members Interests. They are the Conservatives Mr Michael Grylls (Sterling Winthrop), Michael Morris (Upjohn and Reckitt and Colman), Sir Dudley Smith (SmithKline and French) and the Labour MP, Gordon Oakes who advises Riker Laboratories, a drugs subsidiary of the American 3M corporation.

When the debate about the safety of hormone implants in animals reached a critical stage in the EEC, the companies most affected realized that they stood to lose millions of pounds of profits if a ban on the drugs (which has been set for 1988–89) comes into force. The proposed EEC ban may be doubly damaging since it may persuade other countries, including the United States, to reconsider the evidence against hormone implants. As we will see in the next chapter, leading scientists with connections with Roussel Uclaf-Hoechst and IMC were appointed to key positions where they were involved with the outcome of the EEC's final decision. To a lesser extent the major chemical and pharmaceutical companies also spend time and money in an effort to influence the media. Each year the German chemical combines like Hoechst and BASF fly an amphitheatre full of journalists from all over Britain and continental Europe to their headquarters and put them up in comfortable hotels, all expenses paid. The sole purpose of this junket is to enable the enquiring correspondents to hear the company's chairman make an entirely predictable and largely un-newsworthy annual address that could easily have been read on the wire services.

When ICI wanted to demonstrate what strides it was making in selling pesticides and drugs in the United States it flew a group of British journalists (including myself) to New York on Concorde in 1984. The company amply displayed its clout by arranging to have the journalists sit in the cockpit behind the captain of the supersonic craft during the flight. When Concorde landed, executive jets were waiting on the tarmac to ferry this important cargo of creative writers to expensive hotels along the eastern seaboard adjacent to ICI's gleaming installations. To avoid strain on the overworked journalists ICI ended the safari with a pleasure trip to Williamsburg, Virginia, the reconstructed Revolutionary War tourist attraction, before the final executive jet whisked the entourage back to meet the returning Concorde flight to London. Pressure groups like Friends of the Earth and Oxfam, which object to many of the commercial tactics of the drug and chemical industries, normally have to make their protests heard by sending press releases (on recycled paper) by post to journalists' offices.

Lobbying governments, EEC officials and journalists is only one aspect of the influence which chemical and drug companies exert to protect their products and profits. Most of the time they spend their efforts and money in the more mundane task of convincing the customer to buy more of their wares. In agriculture they do this by bombarding the farmer with visits from travelling representatives and by filling the farming journals with advertisements.

Industry sources estimate that the manufacturers of pesticides in Britain now have nearly 6,000 salesmen at their disposal. Around 2,000 are employed directly, and the rest work for agricultural merchant wholesalers and come into the field with an entire arsenal of pesticides supplied by the rival manufacturers.

The pharmaceutical companies employ fewer direct salesmen for their veterinary medicines. The top thirty companies probably employ around forty reps each, or a total of about 1,200. Most of their direct attention is likely to be lavished on doctors because profits from human medicines tend to be higher. Many vets still complain, however, that their surgeries are cluttered with drug salesmen. But the biggest reason why the number of salesmen is lower is that the drug companies are permitted to advertise their prescription medications directly to farmers through the farming press.

Direct advertising to customer/patients of prescription drugs is of course strictly forbidden in human medicine. This ban is applied with equal rigour in the United States. Patients in the care of a doctor should not pressure him to prescribe one wonder cure or another just because its alleged virtues have been hyped up by the ad men.

But on farms just about anything goes. The farming press is littered with advertisements for prescription drugs including those for the human antibiotics, penicillin and tetracycline. One of the biggest companies in the field, American Cyanamid, is a heavy spender in the British farming press. It regularly places full page advertisements in the journal, *Pig Farming*, which urge farmers to use the drug 'Cyfac' because it 'gives pigs the guts to grow'. Cyfac is a mixture of antibiotics including penicillin and tetracycline. Farmers are urged to get their veterinarians to sign prescriptions. Could such heavy promotion lead to excessive antibiotic residues in meat affecting penicillin-sensitive consumers? Could overuse caused by advertising contribute to the pool of disease resistant bacteria which so concern public health officials on both sides of the Atlantic? American Cyanamid's spokesman in Britain, Mr Gary Evans, thought these dangers were grossly exaggerated. He claimed that there was very little evidence to prove that Cyfac and similar drugs were causing the development of resistant strains of bacteria. Any problems that might exist should, he said, be laid at the door of farmers and especially veterinarians.

'As manufacturers we can do absolutely zilch about misuse of our drugs because control is down to the veterinary profession. They are the Luke Skywalkers of this industry. They are the "good guys" and they are responsible for this product which cannot be used without a veterinary prescription.' But at least one of the Cyfac ads made no mention of the need to consult a veterinarian and it failed, even in small print, to say that the drug was obtainable by prescription only. Wasn't it clearly aimed at getting farmers to use Cyfac on a regular, prophylactic basis in their livestock's feed? 'Clearly as manufacturers of the product and being responsible for the bottom line (of profit) we attempt to promote the product, but only within the law,' Mr Evans insisted. But later, in a more reflective mood, he did concede: 'Unfortunately pig production is a highly intensive industry which tends to be surrounded by disease and so by definition if you reduce disease with Cyfac you will give the pigs the guts to grow.'

If Cyfac was used on a regular basis by an unscrupulous farmer who only had profits in mind, would it be safe? He did not believe it would be.

This complacency about advertising is not shared by everyone in the industry. A former director of one large animal drug company was prepared to say: 'I am very uncomfortable about direct advertising of prescription drugs – especially penicillin and tetracycline. They really are abused in animal feed, especially feed given to piglets.' He went on: 'Most of the abuse occurs among the smaller pig farmers who still dominate the industry.' (There are 22,000 pig farmers in Britain.) 'The little guys do a lot of their own pre-mixing of drugs with feed. The approved dose might be one kilogram per ton of feed. But if a farmer thinks he has a bit of a problem he might well chuck in two kilos or even three – and of course, the drug industry reps are not there to discourage him from doing it.

'In my company we always assumed that the lousy farmers used the most of our products because their standards of hygiene were pretty low. We didn't usually bother to send a rep out to see them because they were too small and not worth our while. But we sure did bombard them with ads in the farming press.'

Drug company advertisements can also be very clever in blurring the clear line between high-dose therapeutic antibiotics for genuinely sick animals, as laid down by the Swann Committee report, and the lower dose 'feed' antibiotics bought without prescription for use as growth promoters. By exploiting this grey area of 'preventative treatment' the farmer can be convinced that his profits depend on throwing prescription antibiotics at his livestock . . . even if they seem perfectly well. Take a look at the way that Eli Lilly, the American drug company which also promoted the anti-arthritis drug Opren, chose to promote its prescription-only antibiotic, 'Tylasul', to British pig farmers.

No matter how healthy your pigs appear to be, one thing is certain. They have 'bugs' that cause dysentery, even bugs that are involved in other stress-induced problems. A veritable complexity of bugs, eating at your profits and bringing your business to its knees. Happily, there is one simple solution, Tylasul . . . Talk to your vet about Tylasul now . . .*

The message is clear. Even if your pigs look healthy, are eating well and gaining weight on their regular dose of growth-promoting antibiotics, get your vet to sign them up for an extra drug regime of Tylasul. This kind of hard sell is hardly appropriate for a prescription antibiotic. In the small print of its data sheet Eli Lilly is obliged to warn people who handle the drug to keep it away from their skin. And among the hazards of using Tylasul on animals Eli Lilly has been required to state that it can cause swelling of the mucous membranes of the anus and 'partial protrusion' of the pig's anus itself.

Few farmers ever take the time or trouble to read the fine print of the data sheets which must, by law, accompany the sale of a prescription

* *Farmers' Weekly*, Pig Extra Supplement 11 November 1983, p. 7

drug. Veterinarians, however, should be well versed in these warnings. To counter any concern or squeamishness on the part of the vet, the drug and feed manufacturers have devised ways to overcome his professional caution. Professor Alan Linton of the Bristol Medical School and a member of the government's own Veterinary Products Committee explains: 'Quite often a farmer will ring up his vet and say he is having some cows delivered and he wants some prescription antibiotics because he expects that the animals will be stressed when they arrive.' Professor Linton says that a responsible vet should refuse to recommend antibiotic therapy until he has examined the animals personally and can vouch that they are genuinely 'in his care'. 'But when he gets to the farm he finds that the drug firm has already delivered antibiotics to the farm and he is expected to sign the prescription. He may want to maintain his ethical standards, but he also has to maintain his livelihood.'

The same high pressure techniques – which prey on the farmer's fear of losing profits – have been used to sell growth hormones in Britain.

'Which implant can add an extra 47kg per steer?' – asks Eli Lilly's advert for Compudose, the oestrogen hormone, in bold type across the page. Underneath the copy line is a picture of a weapon that looks like a laser gun from a Star Wars movie. It is actually the injection tool used to squeeze pellets of Compudose under the skin of the cattle's ear. 'One implant promotes extra growth and profits for a full year if required,' says the advert. 'Compudose is a natural choice too. It contains oestradiol 17B, a naturally occurring hormone which means there's no troublesome withdrawal period.'

The phrase 'troublesome withdrawal period' refers to the requirement, not always observed, that farmers must not slaughter animals implanted with the synthetic hormones, Ralgro and Finaplix, for sixty days after these growth boosters have been implanted. This is a good marketing tactic for Eli Lilly. Since it does not have patents for either of the synthetic products, it can make a selling point of the fact that its own 'natural' oestrogen compound can be flowing through the animal's bloodstream to become residues in its meat, in full doses right up to the time of slaughter. The word 'natural' pops up all the time in veterinary drug adverts. Compudose is actually synthesized from other compounds in Lilly's laboratories. It is only natural in the sense that it is supposed to be molecularly identical to the oestrogen normally produced by both male and female cattle.

The makers of the synthetic hormones like to use the word natural too. Ralgro, the synthetic oestrogen, is described to farmers as 'Your natural choice for safe growth promotion'. Once again an impressive looking weapon for insertion of the pellets is pictured below the copy line. Ralgro is actually derived from a fungus that grows on maize and it is never found naturally in any animal species. But, say its makers: 'Ralgro mediates the pituitary gland, thus producing somatrophin, which regulates growth and increases circulating blood sugars. Ralgro stimulates the body's natural production of this hormone (somatrophin), allowing the animal to maximize potential growth by improving food conversion rates, thereby achieving valuable liveweight gains.'

Farmers in Britain have certainly been sold on the use of growth hormones. They believe they are saving £40 million a year by implanting the pellets and have lobbied vociferously to prevent hormones being banned by the EEC. Readers of Ralgro adverts are also reassured that the hormone can be safely used when the cattle are sprayed with pesticides against warble fly, and after they are castrated. Ralgro also complements the use of the antibiotic growth promoter Romensin (another Eli Lilly product) and it can even be used in combination with the synthetic male growth hormone, Finaplix, produced by rival drug company Roussel-Hoechst.

This mixture of chemistry – sprayed on, injected, and fed to cattle – is now accepted by farmers as the norm. It must be stressed again that the chemical companies that produce antibiotics and hormones remain adamant that proper use of their products only speeds livestock weight gain and reduces the price of meat on the dinner table without creating any proven hazard to the consumer. The companies see arrayed against them ill-informed critics, committed environmentalists, enraged vegetarians, and – if the real truth be known – some rather greedy veterinarians who are quick to moan about the hidden dangers of chemical agriculture, but only too eager to enjoy the profits that come from writing prescriptions and dispensing the very drugs that their own professional bodies have expressed grave concerns about.

The British Veterinary Association frequently expresses concern about the hard-sell tactics used by the pharmaceutical companies to get their drugs to the farmers without prior consultation with veterinarians. Yet the BVA decided to launch its own free-sheet newspaper aimed at the farmer called 'You and Your Vet'. All free-sheet newspapers depend entirely on advertising revenue. The first issue of 'You and Your Vet' carried two full-page colour adverts for prescription-only hormones. Both were aimed directly at the farmer. One even offered a free pellet implanting gun . . . for a limited period only. But neither mentioned even in the smallest print that the drug could only be obtained on prescription after authorization from a veterinarian.

The Cyanamid spokesman, Gary Evans, even went so far as to claim that some veterinarians sign prescriptions for his high priced Cyfac antibiotics and then order feed manufacturers to supply a cheaper generic antibiotic substitute. They then pocket the profit derived from the difference in price between the two drugs.

The drug companies characteristically become very exercised whenever their big brand names are threatened. In 1985 the UK Ministry of Agriculture proposed a ban on the sale, supply or importation of carbadox, a feed antibiotic marketed as Fortigro-S by the large U.S. pharmaceutical company, Pfizer. The decision was not taken lightly. The Ministry's scientific advisors had decided that carbadox was a genotoxic carcinogen which should no longer be fed to food-producing animals. Pfizer was furious. Before the ban was threatened Pfizer voluntarily withdrew the drug from the British market. Losing sales in the smaller UK market was not too important. What worried Pfizer was the prospect that a British ban might encourage the U.S. Food and Drug Administration to re-examine its own position.

Carbadox is the second biggest selling feed additive antibiotic in the United States and earned Pfizer sales of $36 million in 1985. In an angry reply Pfizer accused the UK Ministry of Agriculture of ignoring evidence that its product was safe. To call carbadox a genotoxic carcinogen, said Pfizer, represents an absolutist position which 'has been rejected as impractical and scientifically unjustified by responsible regulatory bodies worldwide.'*

The last thing any drug or pesticide company wants is to have the microscope of public concern and scientific scrutiny refocused on any of its big selling products which have been expensively promoted over a number of years to develop customer loyalty. The trouble is that

* *Animal Pharm*, 2 August 1985, p. 2, and 19 July 1985, p. 14

many of the drugs and pesticides now in use were approved for sale decades ago, when techniques to spot cancer dangers were far less advanced.

The UK based chemicals group, May and Baker, discovered to its own cost just how much its own techniques had improved when it re-analysed Ioxynil, a lawn and agricultural weed-killer, that had been on sale for more than fifteen years. The company was forced to re-examine Ioxynil when it applied in 1984 to sell it in a new foreign market, believed to be the United States. The offspring of laboratory rats and rabbits whose mothers were exposed to the weed-killer showed an alarming number of birth defects. To its credit, May and Baker warned the UK Ministry of Agriculture of its findings. The Ministry sat on the results for nine months before taking any action. Then it sent a private letter to May and Baker and other manufacturers advising them that they should cease to make Ioxynil for the domestic gardening market, where pregnant women, trying to remove weeds from their lawns, might be exposed to its tetratogenic (birth-defect causing) properties. But the Ministry failed to call for a similar ban on Ioxynil in agriculture where it is still sprayed on cereals, onions and leeks. Even more disturbingly, the government failed either to warn domestic gardeners of Ioxynil's dangers or to order that it be removed from retail shelves. Industry sources admit that supplies of at least six brands of weed-killer containing Ioxynil but sold under trade names like Actrilawn and New Clovotox would remain on the shelves for up to two years.* Although manufacturing ceased, supplies remained freely on sale. Certainly the government should have cleared the shelves. Consider how many children are born each year with birth defects. All too many. Then think how difficult it has been for mothers who have swallowed suspected drugs during pregnancy to prove in court that their child was maimed by the anti-nausea or sleeping tablet they took. What chance do you think they would have, given all the other factors which *might* have been responsible, of proving that their unborn child was maimed by a garden weed-killer that they happened accidentally to ingest or inhale – even if it were the actual and only cause of the child's terrible affliction? Very little.

The pesticide manufacturers are not in the habit of parading the need for caution. The farming press, which depends heavily on

* The *Guardian*, 25 June 1985, p. 2

advertising revenue for survival, is littered with glossy adverts from the pesticide firms urging farmers to buy their products. Search all day through the back copies and it would be difficult, if not impossible, to find a single advert that warned of the problems of pest resistance. The toxic dangers of pesticides are seldom spelt out – nor is the need to withdraw them from use before harvest to prevent unacceptable residues on food. Instead the adverts prefer to stress that the product has been 'cleared for use' under the Pesticides Safety Precaution Scheme or the formal legislation that replaced it in 1985.

The drug and pesticide corporations take the view that it is their job to promote and sell as much as they can . . . within the law. It is up to the regulatory authorities in Britain and elsewhere to guarantee the safety of these chemicals and to ensure that they are not misused or overused.

6 / The Government Watchdog: Ferocious or Toothless?

Profit, and enthusiasm for their own scientific wizardry, motivate the chemical corporations to push more and more drugs and pesticides into food production. Farmers, eager to maximize their own earnings and yields, have become accomplices in the headlong growth of chemical agriculture.

Against this commercial tide the consumer has only the regulatory authorities of government to protect him and to guarantee that what he buys is clearly labelled and safe to eat. But are the authorities in the United Kingdom and elsewhere doing enough to stop dangerous chemical residues from contaminating our food? The answer, sadly, is that we deserve far better protection than we get and the authorities, too often crippled by complacency or cash cutbacks, are doing much less than they should.

We have only two safeguards: wise regulations drafted by independent and vigilant experts able to spot potential dangers before they occur, and a task force of inspectors equipped to catch offenders who cheat out of ignorance or greed.

The Regulators

Our regulators may be astute, fearless and farsighted, but the concerned consumer can be forgiven for believing otherwise. Nothing engenders suspicion like secrecy. It is a disgraceful fact that many of the decisions on our food safety taken in Whitehall are hidden from proper public scrutiny. Sometimes the excuse is that the Official Secrets Act prevents disclosure. More often senior civil servants, who are trained in the arts of evasion and obfuscation, use their powers to wear down the genuine inquirer by smothering him in rhetoric and red tape before declining, with even-tempered and exasperating politeness, to address the question. Either way, the public is denied the right to know.

Consumers have no right, for example, to examine the toxicology studies performed on rats to test whether a pesticide is likely to cause cancer in man.

Even new guidelines, decreed under the Food and Environment Protection Act 1985, only permit this kind of detailed scrutiny under 'exceptional circumstances', on pesticides which may be approved for use after the legislation's passage. The genuine concern about older pesticides, approved decades ago when toxicology techniques were far more crude, has been conveniently ignored by Whitehall. Even the post-1985 pesticides may only be examined if the government scientists, who have no particular desire to have the public delve into or second guess their decisions, give their express permission.*

All right you say, few of us would understand this complex scientific data anyway. True enough. What we want and need are independent experts, paid at the taxpayers' expense, who can assure us that the chemicals in our food are safe. But the panels of experts who preside on these committees operate in an atmosphere that is far less independent than many might wish.

A high proportion are civil servants who work directly for the Ministry of Agriculture. They do not work for watchdog agencies like the Environmental Protection Agency and the Food and Drug Administration in the United States whose primary job is to protect the consumer. The Ministry of Agriculture, by tradition, atmosphere and express remit, is a sponsoring ministry charged primarily with supporting farmers in their efforts to boost home-grown food production. Consumers who eat the food that farmers grow take a back seat. Second round the committee table are academic scientists. But most of these men have been employed, are being employed or will be employed on research products for the very same drug and pesticide companies who are constantly seeking approval for their new products.

At first glance – and even after a second look – this cosy relationship between government, industry and the farming community is deeply disturbing. The incestuous links are not confined to chemicals and drugs in agriculture. The former Health Minister, Mr Kenneth Clarke, was forced to admit in Parliament in June 1985 that four members of

* The *Guardian*, 13 January 1986

the Committee on the Safety of Medicines (which approves the use of human prescription drugs) were still working as paid consultants to drug companies. But the secrecy that pervades Whitehall permitted him to refuse to name the four, even under cross-examination in the House of Commons.*

Trying to understand who is responsible for controlling chemical residues in food – and what links they have with industry – turns out to be a great deal more complex. An outsider who tries to unscramble the alphabet soup of committees could be forgiven for thinking that the chains of command – and the interrelationships – have been made impenetrable on purpose, especially as all deliberations are covered by the Official Secrets Act. Serious readers with only average memories may wish to use a pen and paper.

The Advisory Committee on Pesticides (MAFF) effectively decides which pesticides are approved for sale, what crops they may be used on, and the permitted dose rate and withdrawal times before harvest. The ACP is also called in to advise when health warnings about pesticides arise.

The Veterinary Products Committee (MAFF) does the same job to approve and reassess the safety and use of veterinary medicines. Both bodies vigorously deny that any of their members have direct links with industry – but again there is no obligation upon Ministers to clear the air. The Liberal MP, Mr Paddy Ashdown has put down numerous written questions in the House of Commons in an attempt to force disclosure of all commercial links within the Advisory Committee on Pesticides.

Once pesticides and veterinary drugs are approved for sale three more committees play a role in determining to what extent they contaminate our food. Here is where the links with industry become obvious.

The Food Advisory Committee is charged with advising Ministers about 'additives, contaminants and other substances which are, or may be present in food or used in its preparation'. The FAC is at least immune to the charge that its industrial links are secret. Four of its thirteen members are directly employed by companies with substantial food interests: Unilever, Reckitt and Colman, Cadbury Schweppes and Grand Metropolitan.

Over at the Department of Health, the *Committee on Toxicity of*

* The *Guardian*, 13 February 1984

Chemicals in Food, Consumer Products and the Environment plays a similar role. And, like the Food Advisory Committee, there are no prohibitions on its members being chosen from people who have had clear links with the industries it is supposed to regulate.

Finally, there is the *Steering Group on Food Surveillance* whose job is 'to collect and review information about the concentrations of contaminants, additives and nutrients in the UK food supply'. It has set up working parties to study drug and pesticide residues and other potential contaminants like food colours and nitrates. Most of its fifteen members work directly for government departments. But two are described as being employed by Unilever and Grand Metropolitan.

In an ideal world there ought to be a clear separation between the watchdogs and the commercial interests they watch on the public's behalf. But in the real world the avenues between academia, industry and government are actually an intricate maze traversed constantly by a relatively small and closely-knit group of men and women. They often perform, interchangeably, all three roles of expert, industrialist and policy maker.

Most of these people see no conflict of interest and believe they can carry out all three jobs with integrity and independence. They see no reason for reproach. It is also true that the government, even if it wanted to, probably could not find enough qualified experts to staff its committees with people who have never had commercial links with industry. Most doctors, veterinarians and academic experts in the safety of drugs and pesticides have at some stage accepted cash from corporations to help fund their own research projects.

This is a far cry from saying that any member of the important food committees has, or would, distort his genuinely held opinions for commercial gain. But people who work closely in and around industry may come to believe and accept that the products produced by these corporations are good for the consumer. If they were cynics and sceptics – critical of corporate behaviour and its unerring righteousness – it is unlikely that they would long keep their jobs or have their research contracts renewed. So it follows that the sort of people who end up on the government's expert committees tend to trust the commercial and scientific process which adds unseen chemicals to our food. Many would find it difficult to imagine that corporate greed could, or would, endanger consumer safety.

Other more radical experts in the medical and academic worlds, who are more mistrustful of the commercial motive, seldom find themselves appointed to the government's watchdog committees in Britain. Within the establishment of Whitehall and the common room they are judged at best to be misguided eccentrics with 'bees in their bonnets' or at worst, to be malevolently motivated Jeremiahs.

One such man is Dr Alastair Hay at the department of chemical pathology at Leeds University Medical School. He has done pioneering work on the toxic properties of many commercial chemicals – sometimes to the considerable annoyance of companies which developed, and now profit by them. His book, *The Chemical Scythe*, combined unimpeachable scientific rigour with a damning indictment of the herbicide 2,4,5-T,* which became known as the notorious Agent Orange, used to defoliate jungles in the Vietnam War. 2,4,5-T and related chemicals are contaminated with potentially lethal dioxins. At Seveso in Italy in 1976 a chemicals plant operated by Hoffmann-La Roche ruptured and sent a similar cloud laden with dioxins into the air injuring thousands of townsfolk. Dr Hay carefully documented the hazards of dioxins and 2,4,5-T, one of the so-called 'dirty dozen' pesticides which pressure groups like Oxfam and Friends of the Earth want to see banned from worldwide use. But 2,4,5-T is still on sale in Britain, approved as safe enough to use under certain conditions by the Ministry of Agriculture.

Dr Hay's expertise is highly valued by government authorities abroad. In 1983 he was invited to serve on a regulatory commission of the U.S. Environmental Protection Agency which looked into the dangers of dioxins in hazardous waste. The following year he was called to testify by an Australian Royal Commission which was investigating the injuries of Australian soldiers who claimed to have been affected by accidental exposure to Agent Orange while serving in the Vietnam War.

Yet in Britain Dr Hay has never been invited to sit on a single government committee, panel or commission where his knowledge and critical views might broaden debate and help ensure that the public is safe from accidental exposure to chemical toxins. A Somerset farmer, Mr Mark Purdy, went to the High Court in London in 1985 in an attempt to prevent the government from forcing him to use organo-

* Hay, Dr Alastair, *The Chemical Scythe: Lessons of 2,4,5-T and Dioxin* (Plenum Press, New York, 1982)

phosphorous pesticides on his cattle to control warble fly. He believed they were dangerous to his family, his cattle and the consumers who bought his milk. The Ministry of Agriculture only requires a delay of six hours between application of the pesticide and milking. Dr Hay combed the scientific literature and found the 'six hour delay rule' to be inadequate. If milk were taken from treated animals long after this period it would contain pesticide residues which exceed the international guidelines laid down by the World Health Organization and the Food and Agriculture Organization of the United Nations.

The Ministry of Agriculture is not eager to have its own expert views challenged. In practice, if not by overt design, the experts chosen by MAFF to sit on its scientific committees usually come from a small, self-selecting circle which excludes the likes of Dr Hay.

This might not be so bad if the government's scientists at least had the opportunity to carry out their own independent tests in laboratories on the drugs and chemicals in food which are passed as safe for human consumption. But they don't. The evidence given to them is almost invariably supplied by the company that wants approval for its product. A parallel might be a court of law which allows the defence counsel to supply the prosecution case against his own client. The animal experiments and other toxicology tests required to bring a single new drug or pesticide to the market can take years and cost many millions of pounds to prepare. Governments have not the money, the lab space or the staff to double check all the results, submitted in huge portfolios of data, by the applicant companies. Much is taken on faith. All the regulatory experts can do is to reread the data, attempt to make sure that the methodology of the tests was sound, and trust to the veracity of the industry scientists who carried out the experiments.

But as the International Bio Test scandal proved (see chapter two), some drugs and pesticides have been approved as safe partly on the basis of bogus trials which were accepted in good faith. Even when safety trials are conducted with scrupulous care, there is no guarantee of safety. Nobody ever lied about most of the drugs and chemicals which have subsequently proved hazardous or lethal. Tests on animals are a very imperfect guide to what the drugs or chemicals will do to humans. There are two reasons for this. The biochemical responses of a homogenous strain of laboratory rats may be very different from the mongrel human population to a particular drug or chemical irritant. Second, all laboratory tests – even long-term studies – are relatively short, conducted over months or at best a few years. None exactly mimic what harm a chemical, swallowed in minute doses, can do to a human over a lifetime. Let us suppose that a particular pesticide, eaten as a residue, produces subtle, but distressing bouts of depression or changes in personality among some people constantly exposed to it. No laboratory study of animals, who cannot easily reveal their emotional distress, would ever sound a warning.

In the case of prescription drugs designed for man, at least some limited clinical trials in human patients are conducted as a double safeguard following animal experiments. But these human trials, which have caught many potentially hazardous medicines before they reach the GPs prescription pad, cannot be used to test the synthetic veterinary hormones like Finaplix or Ralgro – because these drugs were never designed to be used in man. With pesticides the guessing game is even greater. No human volunteer has ever been asked to submit himself in clinical trials to swallowing small pesticide doses to see whether the chemical is safe for the rest of us. Even if such a fictitious volunteer existed, his bravery would undoubtedly be in vain unless he developed a sudden and obvious reaction or died abruptly. The real worry is that tiny doses of pesticides and other agrochemicals can, by slow build up or constant exposure, cause mutagenic or carcinogenic effects to exposed tissue in the same way that smoking causes lung cancer. Of course industry scientists design their laboratory experiments with animals in hopes of detecting just this kind of mutagenic or carcinogenic response to the chemicals they are testing for the market. But there is no clairvoyance, only hindsight. Many potential drugs and pesticides, which have caused tumours to grow in a tiny percentage of the animals tested, have been immediately rejected and discarded long before approval for their use was ever sought. Indeed, since animal tests are an imperfect guide, it is possible that some of these useful products which never came on the market might have proved harmless in man. But it is equally possible that dangerous substances, which showed no adverse effects in animals, have slipped through the net.

Even when the International Bio Test scandal was uncovered, and laboratory studies were proven to be fakes, the regulatory authorities in Britain and the United States were loath to take any precipitate action. Not one drug or pesticide implicated in the scandal was removed from the market in either country. According to Professor Colin Berry, the British government's senior pesticides scientist, withdrawal was not necessary. Why? Because testing is already so secure that fake data from a single source was not enough to warrant any recall. 'The International Bio Test studies did not represent a real gap in the safety data on any of the products licensed in Britain,' he said. Some companies – and he refuses to disclose which – were required to supply new laboratory studies while their products remained on sale. But in all cases, he said, the faked data from International Bio Test only duplicated valid tests already done elsewhere. 'The data package we ask for before we license any product is, frankly, over-elaborate. There is a certain amount of overkill in it and this time the overkill paid off – though I still think that the duplication we ask for to guarantee safety is overdone.'

But could faked data slip through the safety net again? Professor Berry admitted that it could: 'If a chap in the lab decides to feed one of the rats on a box of cream crackers – if he is determined about deceiving people – there is no certain way to find him out.' Britain's Advisory Committee on Pesticides, where Professor Berry leads the investigations, does have the power to retrieve some original evidence from the corporations. It can, for example, demand to see the original microscope slides of animal tissue taken to prove that a particular compound does not cause genetic deformities which may lead to cancer. But the overwhelming bulk of the data scrutinized by the regulatory authorities consists only of columns of numbers on sheets of paper. And Professor Berry admits: 'Ultimately we have to depend on the professionalism of the corporations which submit the data.'

The dilemma for the corporations and consumers alike is that scientific techniques grow ever more precise. Remote links between a chemical and cancer can now, with increasing accuracy, be identified. The whole basis of American approval of drugs, pesticides and cosmetics – anything which the consumer will ingest or apply to his skin – rests on what has become known as the 'Delaney Clause'. In 1958 this amendment was written into the Food, Drug and Cosmetic Act (1938). It required quite specifically that 'no food additive shall be deemed safe if it is found to induce cancer when ingested by man or animal . . .'*

* Schell, Orville, *Modern Meat* (Random House, 1984), p. 193

This broad definition seemed to work pretty well in 1958 when analytical techniques to spot potential carcinogens were pretty crude. Post-war chemicals of all kinds were suddenly pouring on to the market in daily use. But as we now know, many dangerous chemicals, approved as safe, slipped through the net. Scientists of that era were not even able remotely to establish that smoke from cigarettes constantly inhaled onto delicate lung tissue could cause cancer.

But what is the position today? The defenders of the drug and pesticide industry claim that scientific techniques are now so precise, and regulatory demands so strict, that beneficial compounds are kept off the market because somewhere in the bevy of tests an obscure link with cancer is found. Study any substance long enough, they say – from rainwater to the saliva of a mother's kiss – and it will probably prove dangerous in some animal study, somewhere. What do we do then? Ban people from walking in the rain, or mothers from kissing their children? This is the core of the regulatory problem. Has the chemical and drug industry been out of control for a generation – and are our scientific techniques only just catching up? Or are we in danger now of rejecting all new compounds, no matter how useful, because they present a minute and theoretical hazard? The industry defenders have a point. It is frequently said that common table salt and aspirin, for example, are so dangerous that they would never get on the market today if they had to face regulatory scrutiny and approval.

So it is probably true that no drug or pesticide ever comes on to the market these days without some question marks over its safety. At the same time, many products approved as safe twenty years ago when techniques of assessment were crude, remain on the market as potential hazards today, because they have never been submitted to modern scrutiny. There is no 'Delaney Clause' in Britain. Instead, the public must defer only to the judgement of expert committees who weigh the risks and benefits of new drugs and pesticides. And as we have seen, the authorities are loath to allow the public to study the raw data on which their safety assessments are based. Whether we like it or not, we have all become a generation of guinea-pigs exposed by our food, our drugs and the air we breathe to a complexity of chemicals which infiltrate modern life.

Surveillance by the inspectors

Much therefore depends on proper surveillance by our health inspectors to assess whether these chemicals are obviously harmful, or whether their residues, by stealth, are building up in our bodies once they are allowed into the marketplace.

By international standards among developed countries Britain's surveillance record is poor. Germany and the United States, for example, spend far more on food residue testing. If anything, safety standards in Britain have worsened as government cash cutbacks have forced a reduction in the number of inspectors in the field. Laboratory programmes to assess residues in food have also been cut back dramatically.

Yet as we have already seen, many farmers misuse and overuse pesticides, hormones and antibiotics in an effort to boost their yields. What do the authorities do to protect us?

Meat

Each year hundreds of millions of red meat and poultry carcasses pass through UK slaughterhouses. Yet only 300 animals from each species (beef cattle, pigs, lamb and chicken) are systematically tested for antibiotic and hormone drug residues under the government's National Meat Monitoring Programme. All the other animals that troop to their slaughter receive only a cursory visual inspection (if we are lucky) from an environmental health officer or veterinary inspector.

'You cannot look for hormones and antibiotics in the slaughterhouse,' said Mr Clive Wadey, assistant secretary of the Institute of Environmental Health Officers. He believes that real controls on meat are woefully lacking and getting worse. 'We are not going to find out if there is a residue problem in the UK unless we go into a proper, statistically based programme to find the hazard like they do in West Germany.' More than 220,000 meat samples in West Germany were tested for drug and pesticide residues in 1982.* The Germans have been instrumental in pushing the EEC toward a draft directive which would require all member states to apply much more rigorous standards known as 'acceptance sampling' in residue detection. The more costly procedure is being resisted by the Ministry of Agriculture which believes its 300 samples a year method is sufficient.

* House of Lords, European Communities Select Committee, *Examination of Animals and Fresh Meat for the Presence of Residues*. Third Report, 17 December 1985.

But a closer look at present British methods reveals a disturbing picture. The National Meat Monitoring Programme only began in 1980. Although nutritionists have successfully urged us to eat more chicken, not one single poultry carcass was examined until 1985. What dangers are the experts capable of finding once they examine meat in the laboratory? The latest report from the government's Food Surveillance Steering Group admits that when antibiotics are detected the method used 'is unable to quantify or identify the residue present . . . [and] more specific methods are required before meaningful conclusions can be drawn.'*

Mr Wadey believes that the authorities have found the perfect circular argument to justify their complacency. Faced by government cash cutbacks the local authorities (which do most of the detection work) are diverting scarce resources to other, more mundane inspection procedures like checking the water content of hams or fat content of sausages. By doing fewer residue checks they can claim to have found fewer residues. 'And by doing that there is even less chance of finding these hazards in our food,' said Mr Wadey.

By comparison the Americans claim, at least, to have stepped up their residue monitoring in recent years. The veterinary head of the Food and Drug Administration, Dr Lester Crawford, told the Food Editors Conference in Dallas in 1985:

> The quintessential question before us today is what is in the meat. In other words, how are we doing? You would be interested to know that the National Residue Avoidance Programme tests for the presence of seven meat contaminants: (1) chlorinated hydrocarbons; (2) antibiotics; (3) trace elements; (4) nitrogen pesticides; (5) sulfonamides; (6) herbicides; (7) general drugs. You would want to know that 18,603 livestock or red meat samples were analysed in 1984, and that 6,128 poultry samples were tested.†

The Americans cannot afford to be too smug. Although they have banned the highly toxic antibiotic, chloramphenicol, from animal use, they permit unrestricted use of penicillin and tetracycline (two potent common human antibiotics) in animal husbandry. As we have already seen, the pioneering work of the Swann Committee in Britain in 1969 highlighted the danger of antibiotic resistance and attempted to stamp out the uncontrolled use of these drugs. Swann intended this vigilance

* Food Surveillance Paper No. 14, p. 9
† Address to Food Editors Conference, Dallas, Texas, 28 June 1985

to be sustained by a Joint Sub-Committee for Anti-Microbial Substances (JCAMS). This permanent body, drawn from both the Department of Health and the Ministry of Agriculture, was very much part of the Whitehall establishment. But from its creation in 1973 it was refused powers to find out how much antibiotics were being used in British agriculture. And in 1981 it became a victim of government cutbacks and was abolished altogether. Sir James Howie, an original member of the Swann Committee, spoke bitterly about the demise of this important watchdog agency. With unconcealed irony he said: 'Understandably, no doubt, the Ministers preferred the certainty of short term economies to the longer term possibilities of benefit to the health of man and animals.'*

Sir James makes clear that JCAMS refused to shut up – and that is why it was killed off. It wanted its own independent laboratories so that it could check for itself whether virulent strains or resistant strains of bacteria were spreading. It wanted to unlock Whitehall and get some straight answers to some potentially alarming questions about the use of drugs in farming. Whitehall knows how to silence awkward investigators. It abolishes them outright or starves them of funds.

Dr Ray Heitzman's name pops up everywhere in the scientific literature. When the European Commission assembled the best panel it could find on the subject, it summoned Heitzman. But now too, Dr Heitzman's funds to carry on his work in residue detection have been axed completely by the Ministry of Agriculture. What makes this particular story even more disturbing is that Dr Heitzman, in any practical sense, is actually a supporter of the use of hormones as growth boosters in livestock. He has done his scientific sums and reckons that all of the permitted hormones, if administered properly, present no hazard to the consumer of meat. So why has his work at the Institute for Research on Animal Diseases been curtailed? The answer is that Dr Heitzman is extremely good at his job. He has devised techniques to find antibiotic and hormone residues in parts per million – and even per billion – which have previously gone undetected. Surely this is a good thing? No it is not, if you are a government which is eager to tell its trading partners – hands on heart – that it has been unable to find drug residues in its meat. Besides, testing for residues is costly. 'It is a

* Howie, Sir James, *Ten Years on from Swann*, papers at The Association of Veterinarians in Industry Symposium, London, 1981

very expensive business and I don't think the government is eager to put money into it,' said Dr Heitzman. 'The Ministry of Agriculture just does not have the labs or the people to carry on the work.' Dr Heitzman has another serious handicap in the eyes of his Whitehall masters. He is not afraid to speak his views plainly when asked.

Hormones (at least until 1989) and antibiotics are not the only chemicals which contaminate meat. Pesticides, particularly the persistent organochlorines like DDT, pass down through the food chain and become lodged, ultimately in the meat and fat tissue of livestock. An unpublished study on this potential problem was commissioned by the government's working party on pesticide residues (part of the Food Surveillance Steering Group at MAFF). The results, which have been passed to me, showed that more than half (50.9 per cent) of the lamb samples contained 'organochlorine pesticide residues' in excess of the reporting limits set by the government working party. A third of all veal samples, as well as twenty-one per cent of beef samples also exceeded the limits. All other meat and egg samples broke the limits at least six per cent and as frequently as twenty per cent of the time.

Organochlorine pesticides are particularly persistent and pernicious. Many, including DDT, Dieldrin and Lindane, have been linked with cancer. Sheep dipping with Lindane was only banned after the French refused to accept 'contaminated' British lamb. But the Association of Public Analysts, which carried out the tests for the government in June 1985 refused to discuss the contents and importance of its study which it said had been intended for government eyes only. It refused to say which organochlorine pesticides had been discovered in the meat samples or to explain whether the 'reporting levels' of the pesticides constituted a potential health hazard for consumers.

By contrast, the government authorities maintain a posture of calm confidence when they issue their infrequent public reports. The Food Surveillance Steering Group found nothing odd in the fact that no residue testing on poultry had been done prior to 1985 and only casually recommended that it might be a good idea if a few birds were in future tested, 'especially in view of the importance of poultry to the national diet'.

Yet many people may be surprised and alarmed to learn that a staple of their diet – chicken – which is so prone to salmonella infection, escaped any meaningful testing for so long.

Their anxiety will not be relieved by longer quotations from the Food Surveillance Steering Group. Yet this is the body which the government has selected to look after our food safety, and it deserves to speak for itself.

The cancer causing hormone, stilbenes, was finally banned in Britain in 1982. One might expect the Food Surveillance Steering Group to be alarmed if it managed to turn up a single example of this toxic substance in our food when it reported in 1984. But not so. With pride, rather than embarrassment, the Steering Group said: 'The results indicate that there has been a marked decrease in the incidence of positive residues of stilbene growth promoters in bile from food animals since the prohibition of the use of these substances in May 1982.' There is no explanation about why a banned and dangerous substance should find itself in our food at all.

About public concern generally, the Steering Group takes a lofty, almost Olympian, attitude. 'The apparent observation of previously unsuspected chemicals, albeit at low levels, in staple items of diet or the occasional observation of high concentrations of chemicals in particular foodstuffs which are not important parts of the diet can cause concern, particularly in the Press and among the public,' the Steering Group intones. And it concludes reassuringly: 'Whilst such concern may occasionally be well founded, it is often misplaced because the observation has not been put in perspective, or because there has been contamination during sampling or because the accuracy, reproducibility and precision of the observation have not been tested.'*

So we are led to believe that all is well and safe under the control of experts who know better. But buried deeper in the report the Steering Group momentarily reveals its own unease. Occasionally, it says, 'unexpected results' are thrown up by the surveillance programme. Should we be alarmed? Judge for yourself. 'These odd results include, for example, results which appear to be abnormally high or significantly diverge from what is generally accepted as normal. The Steering Group is currently considering how to deal with these aberrant results.'

Sometimes what the authorities consider 'normal' or permissible can be very alarming indeed. Beef cattle in Britain can now be fed on

* Steering Group on Food Surveillance, Paper No. 14, 1984

'chicken litter' – the bedding, faecal droppings, feathers and wasted feed that is swept from poultry pens. Approval to feed beef cattle on this waste was given by the Ministry of Agriculture in May 1985. Chicken litter is apparently high in protein and, not surprisingly, cheap. 'Any ration that can give one kg of weight gain a day for as little as 50p per animal has got to be right economically,' Mr Geoff Walker, a biologist with the Ministry's Agricultural Development and Advisory Service (ADAS) told *Farmers Weekly*.*

Of course, the chicken litter has to be ensiled (turned into fodder correctly). The best method, said ADAS, is to use a thoroughly cleaned muck spreader to mix and fill a clamp with the litter before covering it with a polythene sheet. A cheap carbohydrate, like potatoes or waste material from a breakfast cereal factory, has to be added in small quantity to cut down the acidity of the fermenting mixture. And an official from the Animal Health Office will inspect the final product which should only be fed to beef cattle weighing more than 200 kg (440 pounds). Apparently beef cattle do not much like the taste of chicken litter, and ADAS advises that it should be introduced into their diet gradually.

The Ministry of Agriculture withheld approval of chicken litter for several years because of concerns that salmonella from the droppings might be transmitted to the beef cattle. Now the Ministry reckons it has got the bugs out of the process. 'Daily intake is up to ten kg a head but the feed must be removed seven to fourteen days before slaughter to reduce the risk of residual drugs or hormones that may have been present in the litter feedstuffs.'

The strains of salmonella, which so commonly infect chicken and are transmitted through their droppings, do not normally cross-infect cattle. But at a time when the Public Health Services Laboratory is so concerned about virulent and resistant strains like *Salmonella typhimurium* 204C being transmitted from cattle to man (see chapter three) it is disturbing that the Ministry should appear to put economies in beef production ahead of caution.

When pressed on the growing problem of disease resistant bacteria the Ministry claims to have taken tough new action. As we have already seen, resistant bacteria are likely to spread when farmers circumvent the Swann Committee guidelines by adding high doses of prescription antibiotics to feedstuffs. Their excuse is 'disease prevention', but the

* *Farmers Weekly*, 17 May 1985, p. 20

real motive is often a desire to maximize weight gain of livestock while ignoring the resistance hazard. To prevent this wanton use of drugs the Ministry of Agriculture introduced new regulations in January 1986 which are designed to stop feedstuff manufacturers from including prescription antibiotics in feed without prior written consent from a veterinarian. This is fine in theory, but will it have any clout in practice? Veterinarians will remain under intense pressure to discard their clinical judgement and sign prescriptions for fear of losing the farmer as a customer. And what makes the new regulations even more hollow is the government's refusal to spend any money on enforcement. The job of controlling the feed manufacturers has been handed to the same twenty-odd Pharmaceutical Society Inspectors who are already hopelessly outnumbered and overstretched in their pursuit of black market drug rings and illicit sales of medicines from chemists' shops. To pretend that this handful of men and women can keep adequate check on all farmers and feedstuff manufacturers is to prefer parsimony and cheap public relations to real public protection.

Dr Ray Heitzman, who has now been invited to work in private industry since his public funds have been axed, summed up the failure of the government to provide adequate safeguards: 'I have no doubt that farmers misuse antibiotics and that they do not always observe the proper withdrawal periods. And the Ministry's testing procedures are so inadequate that when abuses occur you can be fairly certain that they won't be found.'

Researchers at the University of Ulster found the same abuse of hormones when they analysed the liver and kidney tissue of four cattle suspected of being improperly implanted with Ralgro and Finaplix. They found residues of the two synthetic hormones which were between 100 and 10,000 times higher than the hormones that would have naturally been found in untreated animals. The results 'give cause for concern', they said, 'especially if the meat products were eaten by sensitive sections of the population. In particular, children or the elderly are most likely to be affected by hormone residues in meat since the natural hormone levels of these subjects are, on average, much lower than that of the general population.'*

* Strain, J. J., et al., *HPLC Analysis of Synthetic Anabolic Residues in Meat, Environmental Health*, Vol. 93, No. 5

Crops

The failure of the government to control adequately the chemical residues in meat should not encourage anyone to become a vegetarian. Surveillance of pesticide use is, if anything, even more haphazard. For a start, the Ministry of Agriculture does not even know with any precision how much pesticide, or what kinds, are applied to fields in Britain every year. Second, it has a very poor idea of how much pesticide residue is left on foods which end up in the shops. The Ministry claims that our 'average daily intake' of pesticides can be extrapolated safely from 'total diet studies' taken from occasional food-basket samples in the laboratory. But Parliament has been unconvinced. The Agriculture Select Committee of the House of Commons began public hearings in February 1986 to probe the effect of pesticides on human health.

Pressure is now growing in the EEC to require member states to adopt and then test for 'maximum residue limits' (MLRs) of pesticides on all food crops. There is nothing revolutionary about this. Countries like Germany have for years had elaborate monitoring procedures that pretty well guarantee MRLs on food in the shops. Nor should there be any real problem in establishing safe limits for each of the common pesticides now in use. An international list of MLRs has been compiled by the Codex Alimentarius Commission, a standing body sponsored jointly by the Food and Agriculture Organization and the World Health Organization of the United Nations.

But the British government has refused to enforce this international guideline and it has fought against MRLs of pesticides being imposed by the EEC. Until 1985 Ministers continued to insist that random checks on pesticide residues, usually carried out by local authorities, were adequate to safeguard public health. As we have already seen, the latest tests by the Association of Public Analysts showed that pesticide residues were found on one third of all fruits and vegetables tested and one in ten samples revealed residues of DDT. This information was not made public by the Ministry of Agriculture, but was disclosed by Friends of the Earth.*

No one outside of Whitehall knows how many random pesticide samples are taken each year. But the number is not thought to exceed a few thousand, and, according to the London Food Commission, the

* Disclosure of some pesticide residues was subsequently made in 1986 by MAFF in the *Report of the Working Party on Pesticide Residues (1982–85)*. Food Surveillance Paper No. 16.

laboratory techniques used are inadequate to detect many of the common but highly toxic pesticides approved for use, especially if the residues appear only in minute quantities.

It is commonly assumed that residues from insecticides present the greatest potential hazard since insecticides, by definition, are designed to be immediately lethal to lower animal species. But a number of the weed-killing herbicides are known or suspected to be carcinogens. No minimum residue of a carcinogen can be proved safe since a single molecule of the compound can trigger development of a tumour.

Even the pesticide manufacturers in Britain, who believe their products to be safe, urged the government to include provision for maximum residue limits on food when the Food and Environment Protection Act 1985 was going through Parliament.

Officials in Whitehall argue privately that the proposed EEC rules are excessive and are likely to be imposed only because the consumer lobby in Europe has triumphed over hard scientific facts. They pointedly remind inquirers that no one in Britain has proveably died from eating pesticide contaminated food.

But why not be certain of safety? The real reason why pesticide residues have not been systematically monitored and legally controlled in Britain is the cost. In a recent television interview the Agriculture Minister, Mrs Peggy Fenner, defensively stated that the government was about to employ sixteen more agricultural inspectors to keep a watch on farmers.* Even this miserly increase failed to materialize. Almost a year later the government was forced to admit (in answer to a Parliamentary question) that it had only increased the inspectorate to 162 and that it had no plans to bolster its strength any further in 1986. And what Mrs Fenner conveniently failed to say was that the number of inspectors had been cut back successively from 176 to 159 in the previous three years. She also did not mention that the amount of money local authorities have to spend on analysing food for chemical residues has also been slashed.

When the new EEC regulations come into force the government will have no choice but to set maximum residue levels of pesticides on crops which are to be exported to other member states. Whitehall officials admit that the anomaly of controlling exported crops while refusing to offer similar protection to British consumers could not be

* '4 What It's Worth', Channel 4, May 1985

tolerated politically. And reluctant plans finally to introduce MRLs on crops consumed at home are now being improvised in Whitehall. Whether the government will commit enough financial resources to ensure adequate monitoring of the scheme remains, however, to be seen. At least in its recent review of food legislation the Ministry of Agriculture admitted that existing regulations under the Food Act 1984 failed to control unintentional contaminants. And it concluded: 'The protection of public health provided by the acts against contamination of food by residues from chemical or microbiological sources needs strengthening.'*

This was a classic piece of Whitehall understatement. Retailers can be prosecuted if they sell food which is 'not fit for human consumption'. But farmers commit no direct offence if they leave excessive residues of drugs and pesticides in the livestock and crops they raise. And without maximum-residue limits set by law how can the courts establish that food has been illegally contaminated with chemicals?

The Supermarkets

This book began with an examination of the chemical residues commonly found in the meat, fruits and vegetables purchased from supermarket shelves. We saw in chapter four that the supermarkets play a large role in determining which drugs and pesticides are used behind the farm-gate. Their heavy involvement in contract farming enables them to stipulate in precise terms what growth-boosting chemicals will be applied to the foods they purchase for our consumption. Given their real control in the market-place, one might expect the major supermarkets to be eager to explain their own role in monitoring the chemicals we are obliged to eat.

Marks and Spencer, which enjoys a high reputation for the quality of its food, refused to answer even written questions about its role in contract farming and about the measures it may take to test all of its meat, fruits and vegetables for antibiotic, hormone and pesticide residues. The store's senior agriculturalist, Mr Peel Holroyd, said that cooperation was impossible because answers 'would involve the disclosure of certain aspects of technology that we regard as commercially confidential'. He seemed unable to fathom that his customers might like to know what chemical residues were likely to be in the M and S

* MAFF, *Review of Food Legislation Consultative Document*, (paras 183–85).

food they bought. The questions I put were designed, apparently, just to sell this book. For Mr Holroyd politely ended his letter: 'May I take the opportunity of thanking you for writing to us and we extend our best wishes for your successful venture.'*

Tesco was sent a similar list of questions about contract farming and its methods of residue testing. The store promised its prompt cooperation. Five months elapsed but repeated phone calls still failed to elicit a single answer. Only after a second letter was sent directly to Tesco chairman, Mr Ian MacLaurin, was a brief and artfully worded response issued. Tesco acknowledged its 'duty to ensure that residues are kept within safe/permissible levels', and it said that all its fresh produce was grown to 'strict specification'. In addition, samples of fresh produce were 'checked regularly'.

Answers such as these can conceal more than they reveal. What does 'checked regularly' actually mean? Tesco ignored a specific request that it be explicit in explaining the methods and the frequency of its residue testing. All the company would say is that its technologists have been working with the Ministry of Agriculture on a research project 'into pesticide residues in fresh foods'.

Tesco's single sentence explanation about controls on hormone and antibiotic residues in the meat it sells was no more enlightening: 'As far as meat is concerned, the company specifies to its suppliers that "meat packed for Tesco Stores Limited must be free of residual growth hormones, or indeed, any chemical contaminants injected, ingested or superficially applied"'. Was Tesco really claiming that it does not accept meat for sale which has come from livestock exposed to hormone implants or antibiotic additives in feedstuffs? For as we already know from Dr Heitzman and other government scientists, some residues of these drugs can be detected in meat even when they have been used in the authorized manner.

Dr Alan Long, the Vegetarian Society Spokesman, managed to get a fuller reply from Tesco when he posed as an ordinary customer (Mr G. Gibson) in a letter written to the company in April 1984. And this time Tesco's story was rather different, indicating that hormones and antibiotics could be used in the production of its meat: 'However, you will appreciate that in some cases it is impossible to trace if such treatments have been used early in the animals' lives,' Tesco's con-

* Letter to the author, 19 August 1985

sumer relations manager, S. Hunt, told the disguised Dr Long. In other words, Tesco appears to behave no differently from other supermarkets – and merely asks suppliers to observe the usual standard withdrawal times before slaughter.

To their credit, both Sainsbury's and Safeway faced up to the issue of chemical and drug residues in their food with less confusion and more candour. Sainsbury said it always attempts to buy on contract so that it can set its own quality control standards. 'The great majority of our foodstuffs have tailor-made specifications covering everything we believe to be relevant to safety, legality and the quality of the product. In all instances of course, foods must conform to Government legislation, or approved codes of practice.'

Fair enough, but what role does Sainsbury's play in monitoring chemical residues once it has stipulated to farmers which antibiotics, hormones and pesticides shall be used on its contracted food purchases? Sainsbury's faced the question:

> Retailers are no more geared up than is Government to continually monitor potential chemical residues in food. At Sainsbury's we have approximately 3,000 own-label food and drink lines alone, from over 700 sources worldwide, distributed in considerable tonnages. Against this background it would be totally impractical in any meaningful way to monitor residues continually.

One of the problems is that food laboratories of the supermarkets simply do not have the sophisticated equipment needed to conduct analysis, in parts per million, of chemical residues in the food they sell.

No doubt the responsible supermarkets make every effort to conform to government regulations. Sainsbury's stressed that it specifically excluded the use of DDT in its fruit and vegetable specifications 'five years before it was banned by MAFF'. But for the most part, the supermarkets do no more than obey the law. In theory a retailer could be prosecuted under the wider provisions of the Food Act for selling foods heavily contaminated with chemical residues. But a senior Ministry of Agriculture official doubted whether a conviction could be secured when the government itself had been so reluctant to set maximum residue levels in law. And even if MRLs came into force, it would be difficult in court to prove that exceeding the regulations constituted a health hazard, he said.

At least one supermarket chain, Safeway, does claim to have introduced some random sampling of its fruits and vegetables. 'Samples are sent off to an analyst's laboratory near Canterbury because the work is so technical we cannot do it ourselves,' said one senior Safeway official. 'But we get no guidance from MAFF. We are just concerned about the amount of chemicals used on our ordinary lines.'

After four years of limited trials Safeway has also begun to market a range of organically grown fruits and vegetables in all of its 125 UK shops (see chapter seven). And the store says that it has tightened the pesticide specifications it lays down for the ordinary commercial growers who continue to supply the vast majority of its produce. 'We certainly are trying to increase awareness among growers and to reduce the arrogant use of agrochemicals,' the senior official said.

The role of the EEC

When maximum residue limits are imposed – setting safer levels of chemicals in our food – the driving force behind them will have been the EEC. This is not surprising. After all, the EEC speaks for the Community, but it does not have to pay the price for greater consumer protection. Individual governments do – a fact which has hardly escaped the notice of the Conservative government, committed as it is to cutting public services and allowing the free market to promote its wares with the minimum of hindrance.

Draft legislation now winding its tortuous way through the European Commission is likely, in the near future, to require Britain to set maximum residue levels on antibiotics and pesticides left in foodstuffs. This is why Ministers have left room in the enabling legislation of the Food and Environment Protection Act 1985 to meet the statutory provisions they expect will be forced upon them by the EEC in relation to pesticide control.

The control of hormone residues has, to some extent, become a separate issue now that the EEC has voted, against the strident opposition of Britain, to ban further use of hormone implants after 1988–89. The proposed ban has been hailed as a great victory for the consumer campaign, led by BEUC in Brussels, which began with the Italian baby food scandal in 1980 (see chapter three). But it is, in fact, a rather irrational triumph for common sense which could still prove hollow if, as is feared, a black market in illicit hormones grows up to beat the official ban. It may be common sense to reduce the amount of chemicals in the food we eat – especially hormone implants which add to the 800,000 tonne mountain of surplus beef stored at huge expense by the EEC. But it is hard to make a rational case for banning hormone implants as a first priority when the hazards from pesticide and antibiotic abuse appear, on the available evidence, to pose a greater threat to the health of consumers.

In the immediate, emotional aftermath of the Italian tragedy which was caused by stilbenes, the European Commission urged that all hormone implants be banned immediately within the Community. Britain and Ireland, which practise castration and had the most to lose economically from the ban, demanded that a scientific panel be assembled before any rash decisions were taken.

Although BEUC continued to organize successful boycotts and protests to support an immediate ban, the Council of Ministers relented and instructed the European Commission to set up an expert committee under Professor Eric Lamming of Nottingham University. Dr Ray Heitzman was also invited to serve. The idea was to push the hormone issue out of the hothouse of public debate and into the cold light of scientific enquiry. It was assumed, naively, that the scientists could reach a reasoned conclusion about hormone implant safety upon which all could agree.

But the Lamming Committee enquiry, which was dogged by delays and in-fighting, was probably doomed from the start. It came to the early conclusion that the three 'natural' hormone implants (laboratory replicas of testosterone, oestrogen and progesterone) presented no health hazard because the residues found in implanted meat were minute in relation to the levels of the same hormones normally produced in the bodies of all consumers, regardless of their age or sex. Most of the argument centred on Ralgro and Finaplix, the two synthetic hormones which, like stilbenes, are never found naturally in any animal species. Unlike stilbenes, Ralgro and Finaplix have 'low oral activity'. They should break down harmlessly when eaten and the Lamming Committee could find no evidence that long term ingestion was dangerous. But Professor Lamming wanted more data. According to him, Ralgro was originally developed but later discarded as a potential birth contraceptive because the oral activity of its synthetic oestrogens was too low. This meant, however, that some of the early toxicology studies on the drug were conducted only on female dogs. There was no point in testing Ralgro on male dogs when it was meant to be a contraceptive pill! As we have seen earlier, the Lamming Committee also had difficulty in giving an absolute assurance about the safety of Finaplix, the synthetic testosterone.

While the Lamming Committee was still unable to issue its final report, consumer activists questioned its scientific impartiality.

Sales of hormones within the EEC are worth in excess of £100 million a year. With that kind of money at stake it is perhaps not surprising that the drug companies would wish to ensure that their voice was heard within the debate. Finaplix is produced by the French drug firm, Roussel, which is itself dominated by its biggest shareholder, Hoechst, Europe's largest drug manufacturer.

Presumably the arguments about hormone safety are waged within the Lamming Committee in the finest spirit of scientific debate. Yet the EEC Farm Commissioner, Frans Andriesson, openly admitted in June 1985 that in its search for scientists of integrity and of the highest calibre the EEC Commission could not exclude those experts who had gained their knowledge by virtue of their links with industry, although, in accordance with EEC Commission policy, these links are, of course, declared.

Of course, no one reasonably expects all scientists in the research community who have ever been paid in some form by corporations to be excluded absolutely from the political debate. Such purity may be fine in principle, but it is quite unrealistic in practice.

But dispassionate observers may be surprised to discover that two of the Lamming Committee scientists have extremely close links with Roussel, the maker of Finaplix. This was admitted by Mr George Powderham, the UK managing director of Roussel, who revealed that one Lamming Committee member, Mr E. Beaulieu, 'is an adviser to the Roussel Group', although not directly in the field of hormones or anabolic steriods. Another Lamming Committee member, Mr A. Rico, was actually used by Roussel as an expert to present the dossiers on Finaplix to the French authorities when the hormone was approved for sale in that country.*

The European Economic and Social Committee, a separate and im-

* Letter from Mr George Powderham to the author, 30 August 1985

portant branch of the EEC framework, also advised the Council of Ministers on hormones and on the wider issue of all chemical residues in foodstuffs. Dr Peter Storie-Pugh, a Cambridge don and former president of the British Veterinary Association, was the Rapporteur for the ESC who advised the Council of Ministers on the whole range of residues, including hormones, in foodstuffs. Dr Storie-Pugh is a highly knowledgeable supporter of hormones and his integrity in searching for a balanced opinion is not in question. But before he drafted the consensus view of the ESC Dr Storie-Pugh availed himself of expert opinion.

The expert scientist he turned to for advice is Dr Martin Terry, a Harvard-educated Texan with a doctorate in veterinary medicine and a PhD in toxicology. Dr Terry is certified by the American Board of Veterinary Toxicology and there are few men in the world who have a better grasp of the complex ways that chemical residence from meat react within the human body.

Dr Terry is also a full-time paid employee of International Minerals and Chemical Corporation, the American firm which owns the other controversial synthetic hormone, Ralgro. He sincerely believes that a consumer would have to eat more than 30,000 pounds of Ralgro-implanted beef in a single day before he would be exposed to a significant health risk – so great, he argues, is the hormone's safety margin. But he also freely admits that he has been sent to Europe by his company with the purpose of defending the drug's safety in Europe, since the proposed EEC ban could also threaten Ralgro's future in the United States. Dr Terry says that the Economic and Social Committee of the EEC is designed explicity as a tripartite body representing the views of industry, trade unions and consumers. And he added: 'Frankly, my participation in the ESC [study group on residues] is largely a thankless task involving an enormous amount of work which appears to offer no compensatory benefits. The ESC is getting some high-quality consulting gratis, as far as I am concerned.'

The consumer groups cannot, however, afford to express too much indignation at the lobbying tactics of the drug companies since they too attempted to place 'friendly' experts near the EEC's decision makers. They also skilfully put pressure on voting blocs within the European Parliament. The agricultural committee of the Parliament, which is dominated by continental farming interests, began to believe that a hormone ban might actually work in its favour. Dairy farmers throughout the EEC had recently been hit by milk quotas. A ban on hormones might put off the awful day when the EEC finally decided to take a cost-cutting shovel to the beef mountain as well.

In October 1985 the European Parliament voted overwhelmingly for an immediate ban on all hormone implants without waiting for the final report of the Lamming Committee which was due in October. Professor Lamming was outraged, calling the decision both 'politically motivated and dangerous'. But the EEC farm commissioner, Frans Andriesson, ignored the furious scientist, suspended his committee before its final report in favour of all five hormone implants could be published, and urged the Council of Ministers to ban the drugs at its next meeting in November.

At first, Britain's Agriculture Secretary, Mr Michael Jopling, let it be known that he would use his veto to block the hormone ban. But his sabre-rattling found no favour with the Prime Minister, Mrs Thatcher, who had just agreed with other heads of state at the EEC summit meeting to restrict vetoes to clear issues of 'vital national importance'. Keeping hormone implants in meat animals hardly fell into this category.

Although Mr Jopling managed to delay the decision for a month, the Council of Ministers voted in December by a majority of nine to one to ban all hormone implants in livestock animals from January 1988. As a sop to Mr Jopling's objections, Britain was given an additional year – until January 1989 – to comply.

Much can happen before the proposed ban comes into force. Shortly after the decision the Ministry of Agriculture said that Britain would attempt to have the ruling overturned in the European Court. Various grounds were put forward: that the majority vote was illegal, that a ban would unfairly disadvantage EEC beef producers because hormone implanted meat would still sneak in from abroad, and finally, that the decision was *ultra vires* because it failed to consider the evidence from the EEC's own scientific experts on the Lamming Committee.

Leaving aside the legal issues, will the ban, if it comes into force, actually work? Any prohibition requires a detection force to catch offenders. Mr Jopling, the Consumers' Association in Britain and the National Farmers' Union all say that a hormone implant ban is unenforceable. Given the present pitiful commitment to surveillance, they are right. Spotting the use of the natural hormones is impossible anyway since a bull with testicles produces more natural testosterone than a castrated steer implanted with the synthetic hormone equivalent. It is possible to detect the use of synthetic hormones, but the British government has already axed the residue testing programme of its leading expert, Dr Ray Heitzman, and it has shown no desire to commit the necessary financial resources to catch offenders using the synthetic (and arguably more dangerous) hormones once the ban is in place.

A hormone ban without proper surveillance could prove a worse health hazard than no ban at all. In some countries like Belgium and Italy, where hormone implant bans were passed well in advance of the EEC ruling, there is strong evidence that a black market in illicit growth boosters is flourishing. In Italy it is reckoned that seventy per cent of all growth-boosting drugs (both antibiotics and hormones) are bought on the black market.* With only 300 cattle inspected for drug residues in Britain each year, it is hard to imagine that farmers who buy hormones on the black market will be deterred by the fear of detection. Nearly four million cattle are slaughtered in the UK each year. Unfortunately, stilbenes have two advantages over the soon-to-be banned hormones. They are cheaper and they leave no tell-tale sign of implant because they can be injected directly into the animal's flesh.

Farmers, of course, are not in the business of harming consumers. But they have a living to make, and like the rest of us, they can be tempted to cut corners and ignore easily-flouted laws when there are handsome profits to be made and the dangerous consequences of their actions seem remote.

After all, is there any certain proof that the abuse of chemicals and drugs in agriculture is harming any of us? The final chapter should help all of us to decide.

* Storie-Pugh, Dr Peter, 'Hormones: A European Problem', *Veterinary Practice*, 1 October 1984.

7 / The Right to Know and the Right to Choose: Critics and Alternatives

'I have no doubt that people die from ingesting pesticide residues and other chemicals in the food they eat,' said Dr Jean Monro, Britain's most controversial expert on diet, allergy and disease. 'No, in scientific terms it has not been proven yet. But we are beginning to document it. The reactions we see are not just allergy, but also acute illness and even death.'

This radical view on the dangers of chemical contaminants in our diet is not, as you might expect, very popular with Dr Monro's mainstream critics within the medical profession. But she can hardly be dismissed as a crank or a charlatan. Her own medical training is classical and her credentials impeccable. She qualified as a doctor at the London Hospital in 1960 and spent her first few years in the traditional disciplines of paediatrics and immunology.

Like many people of devout conviction, Dr Monro began to alter her conventional medical views only when she was faced by a personal crisis. In the 1960s both of her sons began to develop Coeliac disease – the crippling inability to digest wheat and other grains containing gluten. Then her husband was found to be suffering from multiple sclerosis. An urgent desire to heal her own family led Dr Monro to probe the relationship between diet and disease. Today her sons are well, her husband's condition is stabilized and desperate people from all over Britain arrive at her clinic, and unannounced at her Buckinghamshire home, pleading for treatment. Many within the medical establishment continue to view her theories with suspicion. But others, with more open minds, have come to take her views seriously.

Dr Monro believes that between five and twenty per cent of the population are unable to 'detoxify' themselves because they lack the necessary set of enzymes which normally allow us to break down and render harmless the chemicals we swallow. Some of these potential toxins are contained naturally in ordinary foods. Others are put in as additives during food processing. Some of the most toxic are, of course, the unintentional residues from pesticides which are put into our food behind the farm-gate.

The essential enzymes we need to disarm these unseen toxins are called mono-amine oxidases – or MAOs for short. The theory neatly explains why some people can apparently eat packets of additive-laden junk-food, pounds of drug-fed livestock, and bushels of unwashed fruits and vegetables with no observable harm. Their enzyme systems are working efficiently to 'detoxify' them. Yet other people suddenly suffer breathing difficulties, nervous disorders, extreme skin reactions and many other illnesses from apparent exposure to the smallest amount of a 'toxin' they cannot deal with.

'Many people become progressively ill and succumb ultimately to one of the recognized illnesses,' said Dr Monro. 'But the root cause of their disease may be their consumption, in small doses, of a chemical. Yet the death certificate will simply record that, at the end, they died from bronchial pneumonia.'

Many within the medical establishment remain hostile to Dr Monro's thesis that the absorption of hostile chemicals, principally from our diet, can have a profound but little understood effect on the immune and metabolic functions of the body. They do not like to be told that they are merely dealing with the symptoms of disease while she is searching for its cure. But this once yawning gap of suspicion has narrowed, and many more mainstream doctors are now incorporating some of the theories of the 'clinical ecology' movement into their own thinking. The sudden emergence of AIDS (Acquired Immune Deficiency Syndrome) as a modern plague may have played a part in shaking up conventional medical opinion. If one alien intruder – in this case the AIDS virus – can trigger the loss of the immune defences and leave the body open to attack from many diseases, then is it any longer impossible to suppose that the trigger for other illnesses may be chemical invaders?

Britain's most publicized victim of chemical attack has been Sheila Rossall, the former pop singer with Picketywitch, who was variously described by Fleet Street as being 'allergic to the twentieth century' or suffering from 'total allergy syndrome'. Dr Monro played a major role in attempting to rehabilitate her. Ms Rossall's weight had fallen below four stone and she was in a state of total collapse before she was flown out to the Dallas clinic of Dr Bill Rea, who remains Dr Monro's closest colleague. Ms Rossall recovered dramatically as long as she remained in the specially controlled environment of the clinic where the food she ate, the air she breathed and even the articles by her bedside were cleansed of all possible chemical contamination. Sympathy for Ms Rossall diminished, however, when the money donated for her ran out and the Foreign Office grudgingly had to pay £21,000 at the taxpayers' expense to fly her back to Britain in an air ambulance. Dr Monro, who accompanied her on the flight, admits that she has now lost contact with Sheila Rossall and does not know her condition.

Some of Dr Monro's less controversial patients appear to have responded remarkably well and permanently to treatment. Amanda Strang, a thirty-year-old mother and cardiac technician with no previous history of physical or emotional illness, collapsed at her own hospital without explanation in 1980. Her condition deteriorated rapidly, she became allergic to all foods, chemicals and chlorinated drinking water and was near death when she too was sent out to Dr Rea's controlled environment clinic. Once she was 'detoxified' she was gradually reintroduced to many foods and substances and is now able to lead a fairly normal life. But Dr Monro says she must still eat only organically grown fruits and vegetables because any herbicides, even at the lowest level of residue, can trigger another collapse.

The extreme symptoms seen in these patients are fortunately rare. But if the theories of Monro and Rea are correct, then a substantial minority of the 'normal' population with a lesser enzyme deficiency may be suffering from a host of less acute ailments and diseases which are induced by chemical residues – particularly the highly persistent organocholorine pesticides like DDT, Aldrin, Dieldrin and Lindane. Although their use is now either banned or severely restricted, they have built up in the food chain and can remain active for up to twenty years. As we now know, the UK government's working party on pesticide residues was concerned enough about the potential problems of persistence to test ordinary meat and egg samples for contamination in 1985. Although suspiciously high concentrations were found (see chapter six) the analysts refused to discuss their findings with me.

Jean Monro and Bill Rea work closely with Enviro-Health Systems of Dallas, a company which uses the latest laboratory technique, called computer assisted gas-chromatic mass spectrometry, to find the tiniest concentrations of pesticide residues (down to less than one part per billion) in the blood serum of patients. As *The Economist* newspaper reported in 1984, that is the equivalent of 'pin-pointing a grain of salt in a large swimming pool'.* Concentrations this low – even of particularly nasty substances like DDT – might seem innocuous. But as *The Economist* commented in the same article, the organochlorine pesticides are lipid seeking. That means they lodge in fat tissue to reach concentrations 300 times higher than the level found in the blood. Enviro-Health Systems, which is run by Dr John Laseter, is not shy in its diagnosis of what these residues can do. It says: 'Low-dose toxic exposure to common organic chemical compounds is implicated in an ever-broadening range of clinical illnesses, including: hepatitis, lung disease, anaemia, neurological disorders, cancer, reproductive damage, renal dysfunction, immuno-suppression and dermatitis.'†

If this assessment appears extreme, Dr Laseter cannot, on his record, be relegated to the loony fringe. On the contrary, he serves on the science advisory board of the US Environmental Protection Agency and is employed, as a consultant, by more than twenty chemical companies in the United States and abroad. The latter may disagree with his doomwatch diagnosis. After all, if small residues of common pesticides and other chemicals are this virulent, then the very foundations of the modern chemical industry are under threat. But the chemical companies hire Dr Laseter because he is an acknowledged expert in the detective work of assaying trace amounts of lurking compounds. His promotional literature claims 'Laseter was among the first scientists to evaluate moon rock, following U.S. lunar expeditions.' Dr Monro sends her blood serum samples to him because she believes the public analysts in Britain are not well enough equipped to find the trace residues which can trigger disease in many of her patients.

Officially, the British government regards the claims of Monro, Rea and Laseter as scaremongering. In its view, the levels of pesticide residue commonly found on foods in a British supermarket or greengrocer are harmless and do not deserve the epithet 'contamination'. Mr George Trevelyan, a senior civil servant in charge of pesticide control, roundly rebuked Friends of the Earth for using this kind of emotive language. He told FoE that its demand for all British food to be free of

* *The Economist*, 17 November 1984
† Advertising literature which appeared in the *New England Journal of Medicine*

detectable pesticide residues was neither achievable nor desirable.

. . . this would not be achievable even if all pesticide use ceased immediately, in view of the presence of persistent compounds in the environment. Moreover it is an extreme objective, unduly dependent on changing powers of detection or determination, which makes *no direct contribution to health or happiness*. The Government prefers an arrangement which looks at each chemical in turn to assess its toxicity, approves a pattern of use which will be safe for people and the environment, and then aims to ensure that pesticide residues present no risk to public health and are as low as possible commensurate with the effective control of pests and diseases.*

The safe 'pattern of use' referred to by Mr Trevelyan relates to the government's reliance on what is known as the 'total diet study'. If someone were to exist exclusively on a hundredweight sack of potatoes which had been heavily contaminated with fungicides, then MAFF concedes that his total diet might be dangerously contaminated with residues. But in practice most of us eat a great variety of foods, and it is more sensible, using available statistics, to gauge the likely intake of pesticide residues from a typical total diet study.

But critics say this approach is both Olympian and inadequate. 'For a start, the techniques used only allow the government to detect a relative handful of the hundreds of pesticides that have been approved for use in Britain,' said Mr Peter Snell of the London Food Commission.

Chris Rose, the former pesticides campaigner for Friends of the Earth, is even more scathing:

The government has not been doing its monitoring properly and its science is bad. There is genuine ignorance among some of the civil servants who are resisting change, and who are unaware of the toxic properties of some of the substances they are dealing with.

Then there is the close relationship between the chemical industry and government – they are in and out of bed with each other the entire time. You see this at conferences. Whitehall officials and industrialists are obviously comfortable in each other's company, not least because they have spent a lot of time serving on the same committees. The ministry will always bend over backwards in the long run to give industry the benefit of the doubt where safety is concerned.

* Correspondence, Trevelyan to Friends of the Earth, 4 September (italics for emphasis by the author)

In the United States the regulatory agencies like the Environmental Protection Agency and the Food and Drug Administration tend, on the whole, to be less intimate with industry – although a former head of the FDA recently joined the board of a major U.S. drug company. But their ability to find pesticide residues and set safe levels in food comes under similar attack from the leading U.S. environmental pressure group, the Natural Resources Defense Council based in San Francisco. The NRDC claims that U.S. government labs 'can only detect one-third of all the pesticides used on food'. And despite the Delaney Clause which ought to prevent carcinogens being used in food, drug and cosmetic production, the EPA has previously given approval to a number of pesticides, including Dieldrin, DDT, Kelthane and parathion, which are 'suspected or known carcinogens'. Normally, pesticide residues in the U.S. are set at a level 100 times smaller than the dose which shows a 'No Observable Effect Level' (NOEL) in animal studies. This is important because humans can be ten times more sensitive than animals to toxic chemicals and, in addition, some humans are ten times more sensitive than others. But the NRDC claims that the EPA has allowed at least four pesticides to be used with a safety factor of only ten instead of 100.*

The chemical cocktail

Let us draw breath for a moment and assume, as governments insist, that all the pesticides approved for use are safe – that *all* laboratory studies have been done meticulously, and that *all* the short term reactions observed in test animals can be extrapolated to prove that *no* long-term harm has been done or will be done to any human consumer. Even if this is so, each of the chemicals has only been tested in isolation. But as consumers, we are exposed to a constant chemical cocktail of pesticides whose potential dangers, when reacting together in our bodies, have never been explored.

In Britain around 800 different pesticides have been approved for use. Of these, about 400 are commonly used in approximately 4,000 different formulations which may themselves be cocktails of various

* Mott, Laurie and Broad, Martha, *Pesticides in Food: What the Public Needs to Know*, National Resources Defense Council, 15 March 1984

compounds. In addition, Professor Erik Millstone at Sussex University estimates that 3,850 colours, preservatives, flavourings and other additives are permitted for use in food.* How all of these compounds can combine to react in our bodies – or in the bodies of particularly sensitive consumers– is anyone's guess. Studying the interreactions of even a handful of randomly selected pesticides and additives would be an analytical and statistical nightmare. Consequently, health campaigners, who believe that this 'chemical cocktail' is a real but unexplored hazard, have little chance of proving that some consumers may be slowly poisoned by the concoction of compounds they are forced to swallow.

Critics who believe that the misuse of antibiotics is causing visible harm are on much firmer ground. Antibiotic residues in milk and meat do present a real hazard to the admittedly small number of people with extreme allergies to the drugs, especially penicillin. Anaphylactic shock, a potentially lethal reaction, was experienced by a few penicillin-sensitive volunteers who agreed to eat pork from an animal that had been given therapeutic doses of the drug. As we have already seen, farmers in the United States are permitted to use as much penicillin as they like and in Britain the rules to limit its use are seldom enforced and flagrantly ignored. Once again, however, it may be extremely difficult to link a residue with the cause of serious disease. It has been proved that the antibiotic, chloramphenicol, can cause fatal aplastic anaemia and leukaemia. But how could a patient dying from one of these diseases, or his relatives, establish that a residue was the cause when so many unexplained factors can also trigger the onset of illness?

What is not in question, however, is that serious illness and even death is being caused because multi-resistant bacteria, which are bred behind the farm-gate, have spread to man through the misuse of antibiotics. Yet on this issue the UK government has remained strangely silent while it has continued to ignore or to cut the funds of experts hoping to control the outbreak. Sir James Howie, the former deputy chairman of the Swann Committee and medical director of the Public Health Laboratory Service, made no attempt to hide his anger and frustration after the government abolished the Joint Sub-Committee for Anti-Microbial Substances. 'This was perhaps a predictably

* Millstone, Erik, *Food Additives* (Penguin, London, 1986)

philistine reaction in the present economic climate.' But he added: 'The ultimate cost of failure to maintain useful antibiotics is clearly not something to be contemplated.'*

Doctors in Britain are already beginning to see the cost in stark, human terms. Dr Stuart Glover, a Bristol specialist in communicable diseases, recently treated a woman patient in hospital who had contracted *salmonella typhimurium* type 204C – the highly resistant and virulent bacterium which has been rife in cattle. Dr Glover has no doubt that the bug developed because farmers wantonly dose their cattle with therapeutic antibiotics. The abuse only causes the highly promiscuous and adaptable salmonella bacteria to become more resistant and to pass the acquired defence on to other organisms.

'What worries the profession is that the resistance pattern of salmonellas we now see in man is very similar, if not identical to the resistance pattern one sees in animals,' he said. Living in cities away from the countryside is no protection. 'Some of my patients do come from the country,' Dr Glover said, 'but this particular woman actually acquired the 204C infection in hospital. She only survived because we had access to a new and fairly experimental antibiotic.' But developing improved antibiotics cannot solve the problem either. 'Bacteria can learn and pass on resistance to antibiotics they have never even seen. The new drug we tried might have been just as ineffective. My patient is very lucky to be alive today.' Earlier research had suggested that bacteria which carry a multi-resistant gene might at least be 'weighed down' by its defences and therefore less virulent. But Dr Glover added: 'It has been my experience that salmonella which contain a multi-resistant gene can also have enhanced virulence.'

Some experts, including leading members of the British Veterinary Association, believe that the outbreak of resistant and virulent salmonella is cyclical and will naturally run its course. Again Dr Glover disagrees and says the problem is growing worse, not better. 'Most physicians with an interest in infection will see multi-resistant strains of bacteria from time to time. My experience is not unique. People in other parts of Britain have become infected too.' What should be done? 'It is going to take a brave government to ban the use of these antibiotics except for genuine therapeutic use – and how one polices the ban is

* Howie, Sir James, *Ten Years on from Swann*, The Association of Veterinarians in Industry, London, 1981

another matter. I cannot honestly see the veterinarians being policed. If the farmer is not getting what he wants from one veterinarian he will go to another who will supply the drugs. I'm afraid I am pretty cynical about it all.'

One man who maintains an irrepressible zeal in his determination to stamp out the excesses of the meat trade is Dr Alan Long – the seemingly tireless campaigner for the Vegetarian Society. Dr Long leads a curious double life. In his evenings and into the night he works diligently to document the way that veterinarians, farmers and the pharmaceutical companies manage to get more drugs into meat animals than the Swann Committee recommended. By day he works as an organic chemist for a large drug company which also has an important veterinary medicines business including the supply of antibiotics for livestock.

His press releases issued to Fleet Street are difficult to beat for sheer hard-hitting panache. One headlined 'Meat – it's got the lot' began with this paragraph designed to catch even the most bleary journalist's eye:

Steak à la Steroid, Entrecôte Oestrogen, Filet à la Finaplix, McHormone Burgers – Some of the Probable Offers on Menus Today

Dr Long's copyline continues: 'Livestock producers have developed a scandalous disregard for the safety of the consumer, who has no way of identifying these products from "Animal Pharm". Reform is overdue.'

As a committed vegetarian Dr Long crusades against the slaughter of any livestock animals for human consumption. But he is careful to research his attacks thoroughly, and he pulls no punches in criticizing the drug industry which provides his own livelihood.

Modern livestock-farming owes its intensification and mechanization to a deluge of drugs for mass medication ... The drug industry likes the euphemism 'animal health products' for potions plied with little concern for the animal's welfare, but more for the prevalent ills of sick farming, which are stimulated by stress on the livestock and the consequent threats of disease. Doubts over residues are a penalty the consumer pays for 'cheap' meat.

According to Dr Long, hormones and antibiotics are not the only drug residues found in meat. Tenderizing enzymes are injected into animals just before slaughter. 'Although some of the cuts of meat are upgraded in this way, the organs (such as the liver) may be seriously damaged, and the practice entails cruelty.' And he claims that terrified pigs are injected with tranquillizers to calm them just before killing. 'The trade laments the problem of PSE meat (pale, soft and exudative), a typical symptom of ill-treatment of the animal just before it is slaughtered.'

The British Veterinary Association and the Consumers' Association might be expected to endorse many of Dr Long's views, if not always his florid prose style. After all, the BVA exists ostensibly to protect the welfare of animals and the Consumers' Association is supposed to do the same job for us – the consuming public. But both organizations have kept a respectable distance from Dr Long and they take up their stances well within the parapets of the establishment.

Yes, the British Veterinary Association is supposed to defend animals, but its real role, of course, is to defend veterinarians. When those two goals conflict it cannot escape a thorny dilemma. Veterinarians do not have the luxury of working as paid practitioners of the welfare state. They must compete for business among farming clients. They and the BVA cannot afford to antagonize farmers who only want to put more meat on the butcher's hook and may not be too fussy about the antibiotic and hormone drugs they use – and misuse – to maximize their short-term profits. The veterinarians also have to live in harmony with the powerful drug industry. The BVA defends the use of hormone implants and its president, Mr Brian Hoskins, who is head of corporate affairs at Coopers Animal Health Ltd., the ICI-Wellcome veterinary drugs partnership, rejects the idea that the misuse of antibiotics by British farmers is flagrant and widespread. The drug industry would be foolish not to exert its influence within the veterinary profession and there is evidence that this influence is growing. Behind the scenes, however, some senior figures within the BVA are alarmed by the spread of drug resistant bacteria strains from cattle to the human population.

The Consumers' Association finds itself caught in a similar cross-fire on the issue of hormone implants. Two of its senior figures, Daphne Grose and Dr Lesley Yeomans, are frequent advisors to the government on food policy. The Consumers' Association broadly agrees with Ministry opinion that implants of the approved 'natural' hormones, if used correctly, present no scientifically proved hazards to meat and milk consumers.

Dr Yeomans and her colleagues do reserve their doubts about the two synthetic hormones, Ralgro and Finaplix, but the Consumers' Association has never made this issue a significant matter of public concern. And they fear that the proposed EEC ban on the approved hormones would be an 'unenforceable law' which would only lead to black market sales at a time when the government has refused money to catch offenders.

But the Consumers' Association also belongs to, and financially supports, BEUC – the union of European consumer groups – which has led and orchestrated the anti-hormone lobby since 1980. Frenchman Yves Domzalski, who has run BEUC's hormone-ban campaign, despairs at the lack of support he has received from Britain ever since the early days of his successful veal boycott which captured newspaper headlines in the wake of the Italian baby food scandal.

'We have always been careful *never* to say that hormones cause cancer in humans,' said Mr Domzalski. 'What we do stress is that no one can be sure that they do not cause cancer and much more independent research needs to be done. Hormone implants only benefit the pharmaceutical companies that make them. What benefit can there be for the consumer when the EEC already has a huge beef mountain which costs £400 million each year to support?'

Yves Domzalski is a youngish, thoughtful and intense man who has obviously been persuaded by the passion of his own convictions. An unholy alliance of farmers and pharmaceutical companies are lining their own pockets and ignoring risks to consumers. By now he had warmed to his theme and he explained, as we sat in his Brussels office, how growth-boosting drugs also encourage the absurdities of the Common Agricultural Policy.

Calves are not allowed to drink the milk of their mothers because the farmers can sell the liquid milk at a high, subsidized price. Why 'waste' this milk by allowing the calf to drink it when he can be fed instead on cheap milk powder made from the same liquid milk that the CAP buys from the farmer in the first place and cannot dispose of. Besides, the milk powder can be impregnated with antibiotics to make the calf grow even faster. And when his mother gets too old for milking she is implanted with hormones to add as much as sixty pounds to her body weight in a few weeks. Then her slaughtered carcass ends up on the beef mountain . . . it is quite ridiculous.

Within Whitehall, of course, the views of Yves Domzalski are dismissed as emotive and unscientific. Yet he speaks for many consumers who still want to know, with expert opinion divided, why they should be exposed to a potential health risk from growth-boosting chemicals when too much food is already produced and even more chemicals are needed to store and process the unwanted surplus?

At the heart of the matter is the deep unease felt by many people when they are forced to leave decisions affecting their health in the hands of ministers and civil servants whom they may not entirely trust. Unfortunately, 'scientific' facts can be used as props for their own political prejudices. And why should scientific wisdom, which has proved so fallible in the past, suddenly be perfect and clairvoyant now?

This gut anxiety has become irrational but it is entirely appropriate, according to Dr Christopher Wilkinson, the English-born director of the Institute for Comparative and Environmental Toxicology at Cornell University in Ithaca, New York. He believes that the public's fears about chemical carcinogens have been fanned into a phobia by the false claims from politicians and regulators that new scientific techniques can guarantee an absolute margin of safety. 'Despite these advances, however, it is disquieting to realize that we still have only a meagre understanding of the precise mechanisms through which most chemicals exert their toxic effects and as yet have only a very limited capacity to predict the adverse effects of a given chemical in an intact animal, particularly a human animal.'*

As human guinea-pigs we seemingly accept sudden accidental death from chemicals, but not slow insidious destruction.

Apparently, the human psyche is able to accept the fact that under certain circumstances a chemical can cause a rapid demise and, indeed, many widely used domestic and other materials are highly toxic. Sad as they are, the many accidental deaths attributed to such chemicals each year are accepted by society and viewed as something akin to traffic fatalities. What society cannot accept is the possibility, no matter how remote, that long-term, low level exposure to chemicals might ultimately lead to chronic diseases such as cancer or birth defects that are commonly viewed as the ultimate insults to human health.

* All of Dr Wilkinson's remarks are taken from his recent paper: *Risk Assessment and Regulatory Policy*

But how remote is the danger that swallowed chemicals, ingested as unseen and untasted residues, may cause cancer or birth defects in later generations? Dr Wilkinson admits that he does not know, research scientists do not know, and politicians and the public certainly cannot guess.

> We can observe and measure an increased incidence of liver tumours in a population of laboratory rats exposed to 500 parts per million of a given pesticide in its food for two years, but how can we use this information to assess the risk of cancer in humans exposed intermittently to say ten parts per billion of the same pesticide in their drinking water? The answer, of course, is only with a great deal of difficulty, perhaps not at all.

Why do we not know the dangers? First because scientific wisdom is imperfect and is likely to remain so. Second, knowledge costs a lot of time and money – a great deal more than we have spent, or are prepared to spend. To 'prove' that one single chemical, in one single species is – or is not – carcinogenic can cost $1 million and the study takes up to four years to complete. There are 65,000 different commercial chemicals on sale in the United States and the figure cannot be very different in Britain. That would mean spending $65 billion, or nearly four times the cost of the UK Trident missile programme, to give every chemical the best clean bill of health we have. Money well and better spent, perhaps. But the fact is that only eighteen per cent of the drugs, five per cent of food additives and only ten per cent of pesticides currently in use in the United States have been subjected to this kind of complete hazard testing, according to a recent National Academy of Science Report.

So what should we do to protect ourselves? First of all, says Dr Wilkinson, we should stop believing all those politicians and other policy makers who tell us that the risk from swallowing this or that chemical is so many chances in a million or a billion of dying a slow death from cancer or causing birth defects in our unborn children.

'Why do we seem unable to acknowledge the futility of what we are attempting to do, to admit that science is intrinsically unable to answer many of the basic questions we are posing and the numerical estimates of carcinogenic risk are virtually meaningless?'

But at the end of the day Dr Wilkinson remains a reluctant believer. We are all going to die of something, sometime – and the chances of succumbing early to chemical food residue are remote.

'. . . the public must be made fully aware of the fact that at the low levels of exposure usually encountered, the risks associated with the vast majority of synthetic chemicals are truly infinitesimal compared with the many other risks that are faced and accepted as a part of everyday life.'

This may be so. But in many of the other risks – from cigarette smoking, to inoculating our children with whooping cough vaccine, using asbestos sheeting and driving down a motorway – we all have a choice. Unfortunately, most of us cannot choose to avoid the drug and pesticide residues in our food. And none of us can avoid the dangers of contracting a drug resistant infection caused by the misuse of antibiotics in agriculture. Is there any realistic way out of the dilemma?

The alternatives

Organic farming

Eating organically grown food ought to be the easiest way to avoid most of the potential dangers of chemical agriculture. But the organic movement has been plagued with what PR men describe as a twin image problem. Its devotees have been viewed as cranks – worthy crusaders in woolly hats and dung-covered wellies. And the food they grew, while wholesome and nutritious, was also wrinkled, stale, hard-to-find and expensive. So long as the organic movement remained an obscure religion, practised by only a handful of producers and consumers, it was doomed to perpetual failure. A lack of apparent demand meant poor distribution and poor distribution brought only perished produce at premium prices to obscure greengrocers and health-food shops. There appeared no escape from this commercial straitjacket.

Quite suddenly things began to change. In December 1985 Safeway became the first major supermarket chain to stock a sizeable range of organic fruits and vegetables throughout the year at every one of its 125 outlets in the UK. Had Safeway stolen an important lead over its High Street rivals by offering a choice demanded by a growing body of consumers – or has it simply fallen for a fleeting fad? Other supermarkets have also been test marketing a few organic lines in some shops – but this low profile, inexpensive experiment is easy to end when fashion fades.

'The response from our customers has been staggering,' said Mr David Cornish, senior produce buyer with Safeway. 'I am no organic crusader myself. It is immaterial to me what I sell. I am just in the service business to give customers what they want. We have gone national with organics because the customers demand it. I have been in this business twenty-two years – you should see the letters and phone calls we are getting – I have never come across anything else which has caused such a reaction.'

Demand, said Mr Cornish, has come from three kinds of customers.

> First there are the people who are generally concerned about the amount of agrochemicals used in the environment. Then there is the health issue. We have had cancer and allergy clinics thanking us for stocking organics because their patients develop a severe reaction when they eat ordinary food. We have even sent out Red Star Parcels – making up boxes of goodies from our organic growers – for people who don't live anywhere near our shops. And finally, there is what you would call the average shopper who simply wants to know where the flavour in his food has gone. I'm afraid that all of us in supermarkets have been guilty of demanding cosmetic perfection and flavour may have suffered.

Safeway appears also to have taken steps to tighten the pesticides specifications that it demands from ordinary commercial growers and it claims also to have introduced some random residue sampling by sending fruit and vegetables for analysis at an outside laboratory. The big supermarkets have enough muscle in the market place to get many ordinary growers to conform to the regulations they set. But even the supermarkets lack the power to get consistent supplies of organic produce on to their shelves if not enough is being grown.

Until recently less than one per cent of the acreage in Britain had been organically farmed. Changing to organic methods cannot be done on a whim. The waiting period before a field 'detoxifies' and is given the Soil Association's seal of approval can last up to five years. And in extreme cases, where fields have been heavily bombarded with persistent organochlorine pesticides, genuine organic farming may have to wait until the next century.

The man most responsible for putting commercial flair into organic farming – and organic produce on supermarket shelves – is Mr Peter

Segger. His company, Organic Farm Foods Ltd., claims to have set up the first national distribution system for organic foods in the UK with a turnover now exceeding £2 million a year.

'Demand is so strong,' he said, 'that we soaked up all the UK supplies long ago and we are now buying abroad to broaden our seasonal range.' Crops include carrots, parsnips, potatoes, swedes and cabbages from Britain as well as grapes, oranges, lemons, pears and cress from Europe and further afield. 'Safeway has leapt light years ahead of its competitors. They are making money out of it, while getting the credit from consumers – and rightly so.'

Segger and Safeway agree that organic food will remain more expensive. Extra labour and lower yields outweigh the cost savings of not buying fertilizers and pesticides. Price premiums on organic foods range between ten and thirty per cent, but they are likely to fall if production expands, said Mr Cornish, and he added that Safeway is looking into the supply of cheese and fresh meats produced on organic farms without the use of hormones and antibiotics. In the United States such 'additive free' meat is already selectively on sale in at least thirteen states.* In January 1986, the first British venture, 'The Real Meat Company' in Wiltshire, announced the sale of additive free pork, poultry and beef.

If a full range of organic foods were freely available what proportion of customers would pay a premium to buy them? 'Get rid of your old image of organic farming,' said Mr Segger. 'I predict that twenty-five per cent of consumers will be buying organic within the next decade.' Safeway is more cautious. 'Organic produce will never replace our existing lines,' said Mr Cornish, 'but we think it can reach ten per cent of turnover if we have total availability.'

Maybe the popularity of organic food is not a passing fad then. But even the converts concede that it will remain a minority taste. Are there other ways that chemical residues in food may be reduced if not entirely eliminated for all of us?

Integrated pest management

The obvious answer is for farmers to use less pesticides and drugs. All agrochemicals are expensive and the insects, weeds and bacteria they are designed to kill are usually canny enough to learn resistance.

* Associated Press report, 25 September 1985

Sophisticated farmers who want to cut down on residues and resistance have been experimenting with a variety of techniques loosely described as Integrated Pest Management. Some of the solutions are simplicity itself. It is widely known that heavy doses of broad spectrum pesticides kill off not only the target species of insect, but also its own predators. By choosing more selective pesticides farmers can allow these predator insects to survive to do their work for them.

More subtle techniques are also being developed like the use of sex pheromones. These are the chemicals produced by amorous insects that act as sex signals for prospective mates. Fortunately, they can also be synthesized in large quantities in the laboratory and used in traps placed strategically in fields. Males that mistake the traps for the smell of their mates end up being caught or dosed with a small amount of pesticide while the female, at a loss for a partner, cannot mate and propagate.

Farmers can also choose seed varieties which are heartier and more resistant to insect attack, although they may produce lower yields. The trade-off is lower pesticide and fertilizer costs in return for smaller output per acre.

Finally, farmers can resort to the traditional measure of planting crops so as to avoid the peak season of their insect plagues. But the whole impetus of modern farming has been to go for high-input, high-output agriculture. This kind of constant, intensive cropping only makes the pesticide treadmill spin faster as insects from the previous crop are never starved out after harvesting. They merely cross the 'green bridge' and conveniently munch the emerging crop which has been undersown.

The unfortunate drawback of 'integrated pest management' is that it requires commitment and skill, it is not necessarily cheap, and it does not always work. Buying pheromones, for example, can be as expensive as applying pesticides, and they are not much use if the farmer next door does not use them, since pregnant females escaping his pesticides will happily spawn next door.

Second, integrated pest management schemes rely too much on unpredictable Mother Nature whose vagaries have been confounding the farmer for centuries. What is he supposed to do when the predator insect he intentionally avoided killing fails to colonize correctly and eat as instructed the pest that is now devouring his crop? Third, all the schemes really only seek to deal with insects. More than three quarters of the pesticides used in the UK and other developed countries are herbicides and fungicides aimed at competing plant species.

The sad fact is that most farmers prefer the certainties of science, even if they are sold to him, at considerable economic and social cost, by the sales reps of chemical companies. Does science offer the prospect of any safer remedies? Perhaps.

Genetic engineering in agriculture

Before we get into the marvels of biotechnology, let us pause briefly to consider 'microbial pesticides' since they straddle time-honoured pest management and the prospects of genetic engineering. Insects, like the rest of us, catch diseases that can kill them. Around 100 kinds of bacteria, 500 species of fungi and certain strains of viruses naturally infect and destroy insects. Why pollute food and the environment with man's crude and expensive pesticides, if these microbes can be harnessed to do the job with greater safety and precision? At least four companies – Microbial Resources in Berkshire, Monsanto and Abbott in the United States, and Sandoz in Switzerland – now select and propagate batches of these organisms for commercial sale as pesticides to farmers and foresters. Microbial Resources has, for example, successfully sprayed a virus on Forestry Commission pine plantations to treat an attack of sawfly and it has received approval to sell the virus and other organisms in the United States.

Microbial pesticides are not new. The first to be identified, the bacterium *Bacillus Thuringiensis*, was used by the French back in 1938.

Yet today they claim less than one per cent of the £15 billion worldwide pesticide market. Cost is one problem. But the real drawback is the lack of enough natural, attacking species and their failure to do their job in all conditions. Some fungi, for example, pack up completely if the temperature and humidity of the battleground does not suit them.

In the last decade, of course, scientists have learned how to manipulate the genetic material of living things to improve their performance. The development of genetic engineering techniques is already beginning to transform the pharmaceutical and chemical industries. The gene which switches on production of insulin in the human pancreas has been inserted into the genetic material of the bacterium *E.*

coli. The accommodating organism works like a miniature factory to mass-produce the insulin which is purified and has been approved for treatment of diabetes since 1983. Once the concept of inserting useful, foreign genes in host species was grasped, and the techniques perfected, the possibilities of altering the living world became endless. A new gene may be found to improve, for example, the fighting performance of the fungus which hates the cold. Scientists at the University of California at Berkeley have already altered the genetic structure of a bacterium which naturally lives on the leaves of potato plants. They removed the bacterium's gene which unfortunately produces chemicals that encourage ice crystals to form and causes severe frost damage to the potato crop. They wanted to spray their new bacterium in the field in the hope that it would prosper and crowd out the natural, harmful bacteria and give the potato plants frost protection. But the experiment was blocked when environmentalists led by Jeremy Rifkin appealed successfully for a court injunction in 1984. They argued that it was irresponsible to unleash new, genetically manipulated strains of life into the environment until the hazards were fully understood. In November 1985 the U.S. Environmental Protection Agency gave approval for a similar experiment with genetically engineered bacteria to protect strawberry plants from frost damage. But Mr Rifkin again filed suit in a U.S. District Court to prevent the bugs from being unleashed in the environment.

The danger that genetic engineering may disturb the delicate balance of nature is very real. But so too is the damage already being done by the abuse of man-made chemicals and drugs in agriculture.

The seed revolution

Some plants naturally know how to kill insect predators. A variety of the Chrysanthemum produces the poison, pyrethrin, which is more deadly to insects than DDT. Yet this 'insecticide' is almost harmless to man and other species and it does not persist in the food chain. A variety of marigold releases a sulphur compound into the soil which kills nematode worms and a member of the mint family secretes a chemical which gives insects a stomachache.* And other plants release substances which keep down competing weeds.

Can the genetic engineers harness these traits, implant them univer-

* Huxley, Anthony, *The Green Inheritance* (Gaia Books Ltd, 1984), pp. 102–3

sally in other species, and end the need for pesticides? Similarly, some plants, most notably the legumes, are able to fix their own nitrogen for growth. Could we also do away with fertilizers? In sum, is it possible to produce a disease, drought and insect resistant strain of high protein wheat which also fixes its own nitrogen fertilizer and resists intrusion from weeds? Yes. Will it happen tomorrow, next year or even in this century? Perhaps not. The genetic hurdles to overcome are too complex. But the groundwork has been done and the possibilities are not science fiction.

It has not escaped the notice of some observers that the giant pesticide and fertilizer manufacturers in the chemical industry have been going out of their way to buy up seed companies. He who owns the genetic material of today (seeds) will be better able to design the foodstuffs of tomorrow.

You might think the pesticide manufacturers, like ICI, Shell and Ciba-Geigy, who have recently bought seed companies, are trying to put themselves out of business. Unfortunately, these companies may be indulging in a useful strategy of self-protection. If plants can be made resistant to pests, they can also be genetically engineered to tolerate pesticides and so increase the sale of these chemicals worldwide. This view has not been formed by left wing critics of multinational corporations. To the contrary, it is the opinion of the U.S. government's own Office of Technology Assessment taken from its lengthy report on the need for American firms to commercialize biotechnology:

> It should be kept in mind, however, that much of the agricultural research effort is being made by the agricultural chemical industry and this industry may see the early opportunity of developing pesticide-resistant plants rather than undertaking the longer term effort of developing pest-resistant plants . . . It should be noted, though, that increased use of agricultural chemicals could have serious environmental consequences.*

Livestock animals

Injecting livestock animals with sex hormones to increase growth has always been a blunt instrument. What the farmer wants is weight gain from the anabolic properties of the sex hormone. What he would gladly

* *Commercial Biotechnology, An International Analysis*, U.S. Office of Technology Assessment, pp. 177–78

do without is its main sexual function: to promote potency and desire among his 'contented' flock or herd.

In the same way that the insulin producing gene can be removed from the pancreas, so too can pure growth hormone (unrelated to sex hormones) be removed from the genes in pituitary gland and mass produced in microbes. Genetically engineered and produced human growth hormone has just received health authority approval in the United States and Britain for children suffering from dwarfism. By the same route, bovine and porcine growth hormone has been manufactured in the laboratory and is now undergoing extended trials in cattle and pigs.

It is assumed that the proposed EEC ban on the five 'sex' hormones would not bar use of pure growth hormones in Britain or on the Continent.

Monsanto, the U.S. chemicals and drugs firm, is believed to have begun limited trials with pig pituitary growth hormone among farmers in Britain. So far, the results are inconclusive. No adverse effects on meat consumers have been extrapolated. But this 'real' growth hormone may not actually increase the slaughter weight of animals. Instead, early experiments indicate that it may merely cause the livestock animal to lay down more lean meat and less fat. While this may seem desirable, it could impair long-term animal health and productivity since all animals need a proportion of fat to keep them warm and well. Another problem with pituitary hormone is that it cannot be fed to animals because it is destroyed in the gastro-intestinal tract. And doses large enough may be difficult to implant. Another method of boosting growth is being developed by the Meat Research Institute near Bristol. It involves tricking the animal's immune system into knocking out its own somatostatin – a hormone which normally acts as a brake on the production of other growth hormones.*

Of even greater importance may be the development of genetically-engineered vaccines which can give immunity to scours and other virulent farmyard infections that are not too often suppressed by the heavy use of antibiotics which causes disease resistance. Scours is caused by organisms which destroy the ability of cells in the infected animal to retain water. Genetic engineers are now trying to create a scours bacterium which lacks the crucial gene that destroys water

* Stephani Yanchinski, The *Guardian*, 19 July 1984

regulation in the cells of its host.* If these 'harmless' scours bacteria were injected into livestock they might confer immunity to the animals and so prevent both the outbreak of the disease and the 'need' for preventative doses of antibiotics.

The future discoveries and development of science should not, however, be seen as a wonder cure to our current dilemma. For centuries man was plagued by aflatoxins, deadly poisons secreted by moulds which grow most commonly on stored grain. The death toll increased in modern times as our affluence and grain mountains grew. The scourge was finally controlled by a chemical called ethylene dibromide which had the wonderful effect of killing off the deadly moulds. But ethylene dibromide is now believed to be one of the most carcinogenic substances that the chemical laboratory has ever created – and its use has almost universally been withdrawn within the last few years. What next? The latest panacea is irradiation of food. A government advisory committee chaired by Sir Arnold Burgen strongly recommended in April 1986 that the ban on irradiating food in Britain (which had been in force since 1967) should be lifted immediately. Low doses of radiation from cobalt could, it is said, sterilize stored foodstuffs and also eliminate the need for many of the food preserving additives which have given rise to so much health concern. Irradiation is already permitted for some foods in the United States (where ethylene dibromide was the biggest problem) and in other countries. Critics argue that irradiation can damage vitamins and other nutritional elements of food. It could lead to more food poisoning from salmonella if people grew complacent in the mistaken belief that irradiated food would remain germ free. And the danger would always exist that human error or faulty equipment could make our food 'radioactive'.†

At this juncture, solutions to the chemical plague we face are distant and themselves fraught with uncertainty. For the foreseeable future we shall have to live with chemical residues in our food. The immediate way ahead is to find ways to reduce the abuse and to control the present dangers.

* *Commercial Biotechnology, an International Analysis,* U.S. Office of Technology Assessment, pp. 165–66

† Tony Webb, *Food Irradiation in Britain?* The London Food Commission, 1985

Conclusion

Evident control of chemical residues in food, which genuinely reassures the public, will only come when ministers and their obedient civil servants finally abandon their self-protecting resistance to open government.

If Whitehall is so confident that we have nothing to fear from chemical contaminants in our food, why does it close its portcullis to legitimate inquiry? As the Conservative MP, Mr Jonathan Aitken, recently observed: 'We know more about what goes into a pair of socks than about what goes into our food.' It is this climate of secrecy, which breeds suspicion, fear and misunderstanding, that has allowed abuses to occur and which prevents any consensus about the new safeguards which ought reasonably to be adopted.

1. *Secrecy* As a first priority, which bears no real financial cost, we should demand free access to the deliberations and decisions of the expert committees that decide our food policy. This means abolishing the Official Secrets Act restrictions which gag the work of the Food Advisory Committee. It means allowing genuine inquirers to study the raw safety data which the Advisory Committee on Pesticides and the Veterinary Products Committee have used to approve pesticides and veterinary drugs already on sale.

2. *Disclosure* All members of these committees should be required to disclose publicly all their commercial links with the drug and food industries. This must include temporary research contracts, consultancies and direct payroll employment.

These reforms are simple and achievable but they alone do not get to the heart of the problem. A closed, exclusive atmosphere pervades both the Ministry of Agriculture and the Department of Health. By tradition and by statute MAFF exists to sponsor food production and the farming industry. Equally, the Department of Health is charged with nurturing a successful and profitable pharmaceutical industry in Britain. Do we really want our regulators – whose job is to protect the public – to be housed in ministries which work in such cosy alliance with the industries that they, the regulators, are supposed to control? The obvious and radical solution would be to house the Advisory Committee on Pesticides, the Veterinary Products Committee and the Committee on the Safety of Medicines in a separate agency. This is, of course, the American model. The Environmental Protection Agency and the Food and Drug Administration are not perfect watchdogs. But at least they have the clear and unambiguous task of putting the public interest first. And their very existence inspires public confidence.

Realistically, the establishment of separate regulatory agencies in Britain is a long-term goal. But immediate steps can and should be taken to control the use of chemicals in food production.

Pesticides

The Food and Environment Protection Act 1985 already enables ministers to bring in new statutory controls. They should:

1. Set maximum residue levels on food for all pesticides (in line with WHO/FAO guidelines) in advance of any EEC requirement to do so.
2. Withdraw from sale immediately all pesticides where reasonable scientific evidence has linked them with cancer or birth defects in animal studies.
3. Require all approved pesticides to be reregistered for sale every five years to take advantage of improved toxicology techniques.
4. Legitimate investigators must be allowed free access to safety data on all pesticides already in use without prior permission from the Advisory Committee on Pesticides.

Antibiotics

Farmers must be prevented immediately from using sub-therapeutic doses of prescription antibiotics on a regular basis. This practice is in clear breach of the Swann Committee guidelines and has caused drug resistant strains of salmonella and other bacteria to become a real danger to the human population.

1. Direct advertising of prescription-only antibiotics to farmers should be banned, as the Swann Committee recommended in 1969.
2. The British Veterinary Association should put teeth to its concerns and censure members who flagrantly prescribe therapeutic antibiotics for growth boosting use.
3. The antibiotic, chloramphenicol, which has been strongly linked with aplastic anaemia and leukaemia, should be banned from all livestock use.
4. Maximum residue levels (in line with WHO/FAO guidelines) should be made statutory in advance of any EEC requirement to do so.
5. Intensive livestock farmers should be prevented from using fast growing breeds whose durability in modern housing environments is so poor that they must be sustained by frequent or constant doses of therapeutic antibiotics to prevent disease outbreaks.
6. Fish farming methods should be investigated to dispel the suspicion that trout and salmon can only survive intensive, tank conditions if they are fed high doses of antibiotics.

Hormones

1. The British government should drop plans to challenge the proposed EEC ban on hormone implants in the European Court and agree to implement the ban at the earlier date, January 1988.
2. Nevertheless, the report of the Lamming Committee deserves a fair public hearing and it should be published, with or without the consent of the European Commission.
3. Maximum residue levels of hormones in meat (in line with WHO/FAO guidelines) should be made statutory in advance of any EEC requirement to do so.

The British government says it is reluctant to spend more money on monitoring and surveillance of residue levels in food because it claims that standards are already high and there is no clear evidence that the residues present any real threat to consumers. This is a circular argument. So long as the authorities search with inadequate tools for hazards, it is likely that they will not find any. It is disgraceful that government research funds for Dr Ray Heitzman were cut off completely from April 1986. Funds to support this programme should be found immediately. The number of Agricultural Inspectors working for the Health and Safety Executive should also be increased substantially. And it is transparently ridiculous to expect twenty inspectors paid by the Pharmaceutical Society to shoulder the burden of uncovering black market rings and more pedestrian drug abuse throughout the entire country. The government must either contribute substantially to increase their numbers or it should bear the financial burden of establishing its own inspectorate.

A final word. The chemical and food industries, which profit from pesticides, drugs and additives, never tire of reminding us that 'natural' substances from nature can be dangerous too. They are right. Natural plant poisons, like ricin, are more toxic than any pesticide or drug yet devised in the laboratory. But man has had centuries to learn how to select carefully those few plants and animals in nature's store cupboard that he could safely eat – and he has sensibly discarded the rest. Natural selection is a time-tested safeguard. By contrast, we have been forced in just the last few decades to swallow thousands of man-made chemicals whose proven safety remains, unfortunately, a matter of conjecture and controversy.

Index

Acquired Immune Deficiency Syndrome (AIDS) 123
Actrilawn 93
Adams, Stanley 83
additive-free meat 137
additives 14, 102, 123, 128
advertising, food industry 79
aflatoxins 35
agriculture 27–30, 68
Agriculture, Ministry of 26–7, 92–3, 100, 104, 106, 109, 145
 Agricultural Development and Advisory Service (ADAS) 109
 drug control and 48, 96, 110
 food contamination and 113–16
 residue testing by 107, 110
Aimax 38
Aitken, Jonathan 144
Aldicarb 64–5
Aldrin 24, 26, 124
allergy 44, 48, 123–4, 128, 136
American Cyanamid 87–9
anabolic steroids 41
anaphylactic shock 128
Anapolon 84
Andriesson, Frans 118
androgens 41
animal health products 130
animal-rights groups 22–3
antibiotics 15–16, 20, 35–6, 44–9
 advertised in farming press 146
 allergy to 44, 48
 ban on 57–8, 132
 black market in 54–7
 illicit sales of 57, 77
 in livestock feed 14, 16–17, 19, 21, 46
 maximum residue levels 146
 misuse of 46–7, 66, 76–7, 88, 110, 128–31, 135, 146
 overuse of 23, 49, 73–4, 77, 88, 142
 pre-prescription deliveries 90
 prescription 60, 109–10, 146
 regular sub-therapeutic doses 145–6
 residues of 21, 48, 65, 106
 resistance to 17–18, 45–6, 61
 therapeutic 45–7, 73, 129, 143
 see also growth-promoting antibiotics
aplastic anaemia 128
Applebe, Gordon 56
apples 25, 80
Ashdown, Paddy 97
Atkins, Peter 13, 78
avoparcin 46, 71
Avotan 19,71

baby-food, stilbene-contaminated 50–52

Bacillus thuringiensis 139
bacteria, resistant 61–2, 129, 131
BASF 86
Beaulieu, E. 118
Beecham 85
beef 17, 44, 108–9
growth hormones in 14, 18, 39, 41, 43
mountain 117, 132
Belgium, illicit hormone sales in 53–4
Benomyl 25
Berry, Colin 102
BEUC 116–17
Bhopal disaster 31, 34, 84
Bird's Eye 70
birth defects 93, 145
black market in drugs 23, 47, 53–7
blackcurrants 25
bread, wholemeal 26–7
breast milk, pesticide residues in 15
Breckland Farms Ltd 68
British Pharmaceutical Industry Association 72, 85
British Veterinary Association 23, 57, 91–2, 131, 146
Burgen, Sir Arnold 143

Cadbury Schweppes 97
calf milk, medicated 57
calves 16–17, 76–7
cancer 42–3, 136
cancer-inducing additives 102
cancer-linked pesticides 107
capons 39, 53
carbadox, proposed ban on 92
carcinogens 11–12, 14, 24, 127, 132–4, 143, 145
genotoxic 21, 29, 92
Carson, Rachel 33
castration of bulls 39
cattle 17, 41, 46–7, 110
feed 20, 57
cheese-making 15–16
chemical residues, secrecy on 97
chemical revolution 28–30
chicken litter cattle fodder 108–9
chickens 18–19, 52–3, 61, 69–70, 72, 108
children in crop-spraying test 29, 83
chloramphenicol 17, 47–8, 55, 105, 128, 146
chloropicrin 31
chlortetracycline antibiotics 59
cholinesterase inhibition 34
Ciba-Geigy 28–9, 31, 81, 83, 141
Clarke, Kenneth 96–7
clinical ecology movement 123
coccidiosis in chickens 19
coccidiostats 19–20
Codex Alimentarius Commission 111
coeliac disease 122
colourings 20
Compudose 90–91
Conservative government 116
Conservative lobbyists for pharmaceutical companies 86
consumer
choice 135–7
groups 81, 119
preference 72
protection organizations 28
Consumers' Association 120, 131–2
contraceptive, DES as 53
contract farming 67–75, 77–8, 113–14
Cornish, David 136
cows, antibiotic injection into 15
Crawford, Dr Lester 17, 105

crop-spraying test, children in 29, 83
crops, residue testing 111–13
Crouchman, Dr Malcolm 71–2
cucumbers 24
cull cows, hormone implants in 16
Cyfac 87–8, 92

Dalzell, Ian 68
Davidson, Alan 22, 55–6
DDT 13, 15, 20, 23, 24–6, 80, 107, 115, 124–5, 127
  ban on 33–4
  synthesis 32
DES 38, 42, 50–52
Dichlorvos 34
Dieldrin 107, 124, 127
diethylstilboestrol (DES) 38, 42, 50–53
dioxins 15, 99
Domzalski, Yves 132
Dow Chemical company 84
drug revolution in agriculture 28–30
DuPont 83

E. coli infection 61
'E' numbers 14
eggs 19–20
  contract farming control 69–70, 72
Egypt, children sprayed with pesticide in 83
Enviro-Health Systems 124–5
Environmental Protection Agency (US) 14, 28, 36, 99, 127, 140, 145
enzyme deficiency 122–4
enzymes, tenderizing 130
Eskalin 19, 71
ethylene dibromide 35, 143
European Economic Community (EEC)
  ban on antibiotics 132
  bull castration ban 39
  Common Agricultural Policy 30, 67, 132
  hormone black market within 54
  hormone implants ban 14, 86, 91, 120–21; British opposition to 116, 120–21, 146
  maximum residue limits (MRLs) 24, 26, 111–13, 116; British opposition to 111–13
  residue testing standards and 104
European Economic and Social Committee 119
European Parliament, hormone implants banned by 120
Evans, Gary 88, 92

factory farming 19–20, 30, 72–3
Farm Animal Welfare Council 19
farmers 67
  see also contract farming
farming press 87–92, 94, 146
feed industry, use of chemicals dictated by 67–9
feedstuffs, control of 110
Fenner, Peggy 112
fertilizers 141
Finaplix 40–41, 41–2, 90, 110, 117–18, 132
  see also trenbolone acetate
fish farming 20, 146
Flavomycin 19
FMC 70
food 30, 35, 108, 112
  industry 67–9, 79
  safety tests fallible 100–102
  testing residues in 104–7

Food Act (1984) 113
Food Advisory Committee 97, 144
Food and Agriculture Organization 27, 100
Food and Drug Administration (US) 37, 52, 57, 105, 127, 145
Food, Drug and Cosmetic Act (1938) Delaney Clause 102–3, 127
Food and Environment Protection Act (1985) 33, 37, 96, 112, 116, 145
Food Surveillance Steering Group 98, 107–8
Fortigro 21, 29, 74, 92
Freedom of Information Campaign 37
Friends of the Earth 31, 37, 65, 87, 111, 125
fruit 25, 77, 111
fungicides 26, 34–5

Galecron 29, 83
genetic engineering in agriculture 139–140, 141–3
German residue detection 104, 111
Glaxo 85
Glover, Dr Stuart 129
government
  attitude to radical experts 99
  secrecy 95–6, 144
  silencing awkward investigators 106
  view on pesticide residue levels 125–6
grain farmers 67
Grand Metropolitan 97
Greenpeace 84
Griffin, Dr John 85
Grose, Daphne 131
growth hormones 14, 22–3, 37–44, 106, 142
  advertising 90–92
  ban on 18, 40, 63
  deformities caused in animals 41–2
  in livestock 18, 38–40, 132–3
  see also European Economic Community
growth promotion 132–3
  in agriculture 27
  black market 121
  by prescription antibiotics 60, 109–10
growth-promoting antibiotics 19–20, 22, 39, 44, 46, 71
  campaign to ban, in USA 57–60

hamburgers, contaminated 59
Hay, Dr Alastair 99–100
Health Department, pharmaceutical industry and 145
health inspectors 104
Heitzman, Dr Ray 47, 53, 55, 106–7, 110, 117, 121, 147
herbicides 124, 139
  2,4,5-T 15, 34–5, 99
  birth defects caused by 93
  carcinogenic 11–12
hexoestrol 39, 52–3
Hillsdown Holdings 70–72
Hoechst 40–42, 83, 86
Hoffmann-La Roche 83, 99
Holmberg, Dr Scott 58–60
Holroyd, Peel 113–14
hormone implants 43–4
  ban on 86, 91, 98, 120–21
  safety enquiry 117–20
  see also European Economic Community

hormone residues 65, 106
hormones 35, 43
  black market in 53–5, 121
  maximum residue levels 146; *see also* European Economic Community
  misuse of 66, 110, 131
  natural 43, 117, 131–2
  synthetic 40–42, 49, 117–18
  see also growth hormones
Hoskins, Brian 131
Howie, Sir James 106, 128
Hunt, S. 115

Imperial Chemical Industries (ICI) 34, 38, 78–9, 81, 83–4, 141
Inman, Dr Bill 49
insecticides 34–6, 78, 112, 138, 140
inspectors 104, 112, 147
  Pharmaceutical Society 55–6, 77, 110, 147
Institute for Research on Animal Diseases 106
insulin production 139–40
Internatinal Bio Test scandal 36–7, 100, 101–2
International Minerals and Chemical Corporation (IMC) 40–41, 86, 118
Ioxynil 93
Irish Republic, illicit drug sales in 54–6
irradiation of food 143
ITV 36

Joint Sub-Committee for Anti-Microbial Substances (JCAMS) 106, 128–9
Jopling, Michael 120–21

Kelthane 127

labelling of chemical additives 14
lamb 22, 107
Lamming Committee 117–20, 146
Laseter, Dr John 125
lettuce 24
leukaemia 128, 146
Lilly, Eli 89
lincomycin 46
Lindane 15, 24–5, 26, 107, 124
Linton, Professor Alan 61–2, 90
lipid-seeking pesticides 15, 125
lobbying 85–7, 119–20
Lockwoods and Smedleys 70–72
Lomotil 83
Long, Dr Alan 114–15, 130

MacLaurin, Ian 114
Major, Chris 79
Marks and Spencer 71, 78, 113–14
May and Baker 93
meat 30, 45, 107, 137
  testing for residues 104–10
Meat Research Institute 142
Medawar, Charles 83
medical monitoring, need for 35
Medicine Act (1970) 46
methyl isocyanate (MIC) 31
Microbial Resources 139
Milan, sexual precocity in 50–52
milk 15, 48, 100, 132
Milk Marketing Board 15, 69
Millstone, Professor Erik 128
mono-amine oxidases (MAOs) 123
Monro, Dr Jean 65, 122–4, 125
Monsanto 37, 142
  and Abbott 139
moulds 34–5, 143

Müller, Paul 31
mutton 22

National Farmers' Union 69, 120
National Meat Monitoring Programme 105
Natural Resources Defence Council (NRDC) 58–9, 127
nerve gases 31, 34
New Clovotox 93
nitrofuran 47
nitrogen-fixing plants 141
Nitrovit 71
Nutrikem 71

O'Brien, Dr Thomas 58–60
oestradiol 40–41, 90
oestrogens 41, 53, 90
offal, high antibiotic levels in 21
Office of Technology Assessment (US) 141
Official Secrets Act 144
oil companies 32
Opren 49, 89
Organic Farm Foods Ltd 137
organic produce 116, 135–7
organochlorines 15, 20, 24, 32, 34, 80, 107, 124–5, 136
organophosphorous pesticides 31–2, 34–5
Ortquist, Melissa 84
Oxfam 84, 87

paraquat 34, 37
parathion 127
pastures 15
Pauls' 68
pears 25
penicillin 44–5, 57, 59, 88, 105, 128
pest management, integrated 137–9
pesticide residues 23–7, 32, 37–8, 78, 79–80, 100, 124–5
  Agriculture Ministry's view of 14
  British rules on 65
  checks on 72, 111–13, 126
  deaths from 122
  long-term effect 65, 101
  quantities of 14
  in sludge 15
  toxic 123, 125
  in watermelons 64–6
pesticides 23, 30–37, 49, 94, 99, 107, 128
  cancer-linked 107, 127
  interreactions 127–8
  lipid-seeking 15, 125
  manufacturers of, buying up seed companies 141
  microbial 139
  misuse of 13, 66
  organophosphorous 99–100
  reregistering 145
  research results secret 96
  sales of 32–3, 87
  selective 138
  testing 36–7, 49, 78
  see also organochlorines
Pesticides Advisory Committee 36, 97, 102, 145
Pesticides Safety Precaution Scheme 35
Pfizer pharmaceutical company (USA) 21, 29, 74, 92
pharmaceutical companies 81–3, 85, 87, 96–7
  advertising 87–92
  self-defence 81–3, 84–7, 91, 103
Pharmaceutical Society inspectors 55–6, 77, 110, 147
pheromones 138

phosgene 31
pig pituitary growth hormone 142
pigs 38, 47, 74–5
contract farming control 68–9, 71
pituitary hormone 142
Planta, Dr Louis v. 28, 81
plastic anaemia 146
plums 25
pollution, water 20
potatoes 26
poultry 18–20, 69
residue testing 105, 107–8
see also chickens
Powderham, George 118
prescription drugs 54–7, 87–90, 92
progestagens 41
progesterone 40–41
Public Analysts, Association of 23, 107, 111
Public Health Service Laboratory 47
Puerto Rico, sexual precocity in 52–3
Purdy, Mark 99–100
pyrethrin 140

Ralgro 10, 40–43, 90–91, 117–19, 132
see also zeranol
Rea, Dr Bill 124
Real Meat Company 137
Reckitt and Colman 97
regulatory authorities 95–103
favouring industry 98, 126
need for separate agency 145
reproduction, impaired by growth hormones 42
residue levels
exceeding World Health Organization guidelines 16
maximum 111–13, 115–16, 145–6; *see also* European Economic Community
residue testing, supermarkets and 113–16
residues, secrecy on 144
ricin 147
Rico, A. 118
Rifkin, Jeremy 140
Rollins, Bob 64
Rose, Chris 126
Rossall, Sheila 123–4
Roundup 37
Roussel Uclaf-Hoechst 86, 91, 118
Rowe, Dr Bernard 18, 47, 62

Saenz de Rodriguez, Dr Carmen A. 52
safety 104, 144–5, 147
Safety of Medicines, Committee on the 97, 145
Safeway 115–16, 135–7
Sainsbury's 68, 71, 78, 115
salmonella 17–19, 23, 47, 77, 109
antibiotics and 45–7, 58–60, 62–3
casualties from 58, 60, 62
resistant strains 45–7, 109, 129
Sandoz 139
scours 17, 47, 74, 142–3
Searle, G. D. 83
seed companies 141
Segger, Peter 136–7
sewage sludge 15
sex pheromones 138
sexual deformities, growth hormones and 18, 41–3
sexual precocity, stilbenes and 50–53
sheep, growth hormones and 38
Shell 141

Shoentall, Dr Gina 66
SmithKline 82
smuggling black market drugs 54–7
Snell, Peter 126
soil, detoxification of 136
Soil Association 23
Solomon, Harry 70
somatostatin 142
Speight, David 71
steroids, anabolic 41
stilbenes 38–9, 42–3, 53, 108
illicit supplies 50–51
premature sexual development and 50–53
Storie-Pugh, Dr Peter 119
Strang, Amanda 124
strawberries 25
supermarkets 13, 77–8, 113–16
farm output control by 68–9
organic produce and 135–7
Swann Committee 45–6, 57, 89, 105–6, 109, 130, 146

Tagamet 82
tartrazine 20
tecnazene 26
tenderizing enzymes 130–31
Terry, Dr Martin 119
Tesco 114–15
testosterone 40–41, 117, 120
tetracycline 44–5, 57, 59, 88, 105
thalidomide 42, 49
Thatcher, Margaret 120
thelarche 52–3
Third World 28, 81, 83–4
Thompson, David 70
Threlfall, Dr John 63
tomatoes 24
toxic chemicals, long-term low-level exposure to 133–5
Toxicity of Chemicals in Food, Consumer Products and the Environment Committee 97–8
tranquillizers, pre-slaughter 131
trenbolone acetate 40–42
see also Finaplix
Trevelyan, George 125–6
Triazophos 34
Tylasul 89

Unilever 97
Union Carbide 31, 64, 84
United Kingdom 54–7, 66
United States 64–6, 105
salmonella infection in 57–63

vaccines, genetically engineered 142–3
veal 39, 41, 50–52, 107
vegetables 25–6, 72, 80
testing 77, 111
Vegetarian Society 22
veterinarians 23, 48, 56–7, 75, 82, 92
Veterinary Products Committee 48, 61, 97, 145
virginiamycin 46, 71

Wadey, Clive 104–5
Wakefield hospital salmonella disaster 60–61
Walker, Geoff 109
Walton, Dr John 72
Walton, John (farmer) 73–4
warble fly control 100
weed-killers: *see* herbicides
wheat, wholemeal 26
wheatgerm, oestrogen in 43
Wilkinson, Dr Christopher 133–5
Wingfield, Joy 77
Wishart, Dr David 57
World Council of Churches 28, 81

World Health Organization 27, 100
Wyer, Charles Richard 56

Yeomans, Dr Lesley 131–2

zeranol 40–41
  see also Ralgro
zinc bacitracin 71
Zuckerman committee 33

MORE ABOUT PENGUINS, PELICANS, PEREGRINES AND PUFFINS

For further information about books available from Penguins please write to Dept EP, Penguin Books Ltd, Harmondsworth, Middlesex UB7 ODA.

In the U.S.A.: For a complete list of books available from Penguins in the United States write to Dept DG, Penguin Books, 299 Murray Hill Parkway, East Rutherford, New Jersey 07073.

In Canada: For a complete list of books available from Penguins in Canada write to Penguin Books Canada Ltd, 2801 John Street, Markham, Ontario L3R 1B4.

In Australia: For a complete list of books available from Penguins in Australia write to the Marketing Department, Penguin Books Australia Ltd, P.O. Box 257, Ringwood, Victoria 3134.

In New Zealand: For a complete list of books available from Penguins in New Zealand write to the Marketing Department, Penguin Books (N.Z.) Ltd, Private Bag, Takapuna, Auckland 9.

In India: For a complete list of books available from Penguins in India write to Penguin Overseas Ltd, 706 Eros Apartments, 56 Nehru Place, New Delhi 110019.